# Conference Proceedings of the Society for Experimental Mechanics Series

*Series Editor*

Kristin B. Zimmerman, Ph.D.
Society for Experimental Mechanics, Inc.,
Bethel, CT, USA

The Conference Proceedings of the Society for Experimental Mechanics Series presents early findings and case studies from a wide range of fundamental and applied work across the broad range of fields that comprise Experimental Mechanics. Series volumes follow the principle tracks or focus topics featured in each of the Society's two annual conferences: IMAC, A Conference and Exposition on Structural Dynamics, and the Society's Annual Conference & Exposition and will address critical areas of interest to researchers and design engineers working in all areas of Structural Dynamics, Solid Mechanics and Materials Research.

More information about this series at http://www.springer.com/series/8922

Ming-Tzer Lin • Cesar Sciammarella • Horacio D. Espinosa
Cosme Furlong • Luciano Lamberti • Phillip Reu • Michael Sutton
Chi-Hung Hwang
Editors

# Advancements in Optical Methods & Digital Image Correlation in Experimental Mechanics, Volume 3

Proceedings of the 2019 Annual Conference on Experimental and Applied Mechanics

*Editors*
Ming-Tzer Lin
Graduate Institute of Precision Engineering
National Chung Hsing University
Taichung, Taiwan

Horacio D. Espinosa
Mechanical Engineering
Northwestern University
Evanston, IL, USA

Luciano Lamberti
Politecnico di Bari
Bari, Italy

Michael Sutton
Department of Mechanical Engineering
University of South Carolina
Columbia, SC, USA

Cesar Sciammarella
Illinois Institute of Technology
Chicago, IL, USA

Cosme Furlong
WPI-ME/CHSLT
Worcester Polytechnic Institute
Worcester, MA, USA

Phillip Reu
Sandia National Laboratories
Albuquerque, NM, USA

Chi-Hung Hwang
Taiwan Instrument Technology Institute, NARLabs
Hsinchu, Taiwan

ISSN 2191-5644     ISSN 2191-5652  (electronic)
Conference Proceedings of the Society for Experimental Mechanics Series
ISBN 978-3-030-30008-1     ISBN 978-3-030-30009-8  (eBook)
https://doi.org/10.1007/978-3-030-30009-8

© Society for Experimental Mechanics, Inc. 2020

This work is subject to copyright. All rights are reserved by the Publisher, whether the whole or part of the material is concerned, specifically the rights of translation, reprinting, reuse of illustrations, recitation, broadcasting, reproduction on microfilms or in any other physical way, and transmission or information storage and retrieval, electronic adaptation, computer software, or by similar or dissimilar methodology now known or hereafter developed.
The use of general descriptive names, registered names, trademarks, service marks, etc. in this publication does not imply, even in the absence of a specific statement, that such names are exempt from the relevant protective laws and regulations and therefore free for general use.
The publisher, the authors, and the editors are safe to assume that the advice and information in this book are believed to be true and accurate at the date of publication. Neither the publisher nor the authors or the editors give a warranty, expressed or implied, with respect to the material contained herein or for any errors or omissions that may have been made. The publisher remains neutral with regard to jurisdictional claims in published maps and institutional affiliations.

This Springer imprint is published by the registered company Springer Nature Switzerland AG.
The registered company address is: Gewerbestrasse 11, 6330 Cham, Switzerland

# Preface

*Advancements in Optical Methods & Digital Image Correlation in Experimental Mechanics* represents one of six volumes of technical papers presented at the 2019 SEM Annual Conference and Exposition on Experimental and Applied Mechanics organized by the Society for Experimental Mechanics and held in Reno, NV, June 3–6, 2019. The complete proceedings also include volumes on *Dynamic Behavior of Materials; Challenges in Mechanics of Time-Dependent Materials, Fracture, Fatigue, Failure and Damage Evolution; Mechanics of Biological Systems and Materials & Micro- and Nanomechanics; Mechanics of Composite, Hybrid and Multifunctional Materials; and Residual Stress, Thermomechanics & Infrared Imaging and Inverse Problems.*

Each collection presents early findings from experimental and computational investigations on an important area within Experimental Mechanics, Optical Methods and Digital Image Correlation (DIC) being important areas.

With the advancement in imaging instrumentation, lighting resources, computational power, and data storage, optical methods have gained wide applications across the experimental mechanics society during the past decades. These methods have been applied for measurements over a wide range of spatial domain and temporal resolution. Optical methods have utilized a full range of wavelengths from X-ray to visible lights and infrared. They have been developed not only to make two-dimensional and three-dimensional deformation measurements on the surface but also to make volumetric measurements throughout the interior of a material body.

The area of DIC has been an integral track within the SEM Annual Conference spearheaded by Professor Michael Sutton from the University of South Carolina. The contributed papers within this section of the volume span technical aspects of DIC.

The conference organizers thank the authors, presenters, and session chairs for their participation, support, and contribution to this very exciting area of experimental mechanics.

| | |
|---|---|
| Taichung, Taiwan | Ming-Tzer Lin |
| Chicago, IL, USA | Cesar Sciammarella |
| Evanston, IL, USA | Horacio D. Espinosa |
| Worcester, MA, USA | Cosme Furlong |
| Bari, Italy | Luciano Lamberti |
| Albuquerque, NM, USA | Phillip Reu |
| Columbia, SC, USA | Michael Sutton |
| Hsinchu, Taiwan | C.-H. Hwang |

# Contents

| | | |
|---|---|---|
| 1 | **Digital Projection Speckle Technique for Fringe Generation** ................................................................. <br> Austin Giordano, Andrew Nwuba, and Fu-Pen Chiang | 1 |
| 2 | **Quantifying Wrinkling During Tow Placement on Curvilinear Paths** ..................................................... <br> Sreehari Rajan, Michael A. Sutton, Roudy Wehbe, Brian Tatting, Zafer Gürdal, Addis Kidane, and Ramy Harik | 9 |
| 3 | **Experimental Mechanics, Tool to Verify Continuum Mechanics Predictions** ....................................... <br> C. A. Sciammarella, L. Lamberti, and F. M. Sciammarella | 13 |
| 4 | **Study the Deformation of Solid Cylindrical Specimens Under Torsion Using 360° DIC** ...................... <br> Helena Jin, Wei-Yang Lu, Jay Foulk, and Jakob Ostien | 33 |
| 5 | **Multiscale XCT Scans to Study Damage Mechanism in Syntactic Foam** ............................................ <br> Helena Jin, Brendan Croom, Bernice Mills, Xiaodong Li, Jay Carroll, Kevin Long, and Judith Brown | 37 |
| 6 | **An Investigation of Digital Image Correlation for Earth Materials** ..................................................... <br> Nutan Shukla and Manoj Kumar Mishra | 41 |
| 7 | **Dynamics of Deformation-to-Fracture Transition Based on Wave Theory** ......................................... <br> Sanichiro Yoshida, David R. Didie, Tomohiro Sasaki, Shun Ashina, and Shun Takahashi | 47 |
| 8 | **Fatigue Monitoring of a Dented Pipeline Specimen Using Infrared Thermography, DIC and Fiber Optic Strain Gages** .............................................................................................................................. <br> J. L. F. Freire, V. E. L. Paiva, G. L. G. Gonzáles, R. D. Vieira, J. L. C. Diniz, A. S. Ribeiro, and A. L. F. S. Almeida | 57 |
| 9 | **Development of Optical Technique For Measuring Kinematic Fields in Presence of Cracks, FIB-SEM-DIC** ........................................................................................................................................ <br> Y. Mammadi, A. Joseph, A. Joulain, J. Bonneville, C. Tromas, S. Hedan, and V. Valle | 67 |
| 10 | **DIC Determination of SIF in Orthotropic Composite** ............................................................................ <br> N. S. Fatima and R. E. Rowlands | 75 |
| 11 | **Determining In-Plane Displacement by Combining DIC Method and Plenoptic Camera Built-In Focal-Distance Change Function** ............................................................................................................ <br> Chi-Hung Hwang, Wei-Chung Wang, Shou-Hsueh Wang, Rui-Cian Weng, Chih-Yen Chen, and Yu-Chieh Chen | 79 |
| 12 | **Identification of Interparticle Contacts in Granular Media Using Mechanoluminescent Material** ............ <br> Pawarut Jongchansitto, Damien Boyer, Itthichai Preechawuttipong, and Xavier Balandraud | 87 |
| 13 | **Colour Transfer in Twelve Fringe Photoelasticity (TFP)** ........................................................................ <br> Sachin Sasikumar and K. Ramesh | 93 |
| 14 | **Infrared Deflectometry** ........................................................................................................................... <br> H. Toniuc and F. Pierron | 97 |
| 15 | **Real-Time Shadow Moiré Measurement by Two Light Sources** ............................................................. <br> Fa-Yen Cheng, Terry Yuan-Fang Chen, Chia-Cheng Lee, and Ming-Tzer Lin | 101 |

**16 Study of MRI Compatible Piezoelectric Motors by Finite Element Modeling and High-Speed Digital Holography** .................................................................................................................. 105
Paulo A. Carvalho, Haimi Tang, Payam Razavi, Koohyar Pooladvand, Westly C. Castro, Katie Y. Gandomi, Zhanyue Zhao, Christopher J. Nycz, Cosme Furlong, and Gregory S. Fischer

**17 Digital Volume Correlation: Progress and Challenges** ............................................................. 113
Ante Buljac, Clément Jailin, Arturo Mendoza, Jan Neggers, Thibault Taillandier-Thomas, Amine Bouterf, Benjamin Smaniotto, François Hild, and Stéphane Roux

**18 Development of 3D Shape Measurement Device Using Feature Quantity Type Whole-Space Tabulation Method** .................................................................................................................. 117
Motoharu Fujigaki, Yoshiyuki Kusunoki, and Hideyuki Tanaka

**19 Temporal Phase Unwrapping for High-Speed Holographic Shape Measurements of Geometrically Discontinuous Objects** .................................................................................................. 121
Haimi Tang, Payam Razavi, John J. Rosowski, Jeffrey T. Cheng, and Cosme Furlong

**20 Projection-Based Measurement and Identification** ................................................................. 125
Clément Jailin, Ante Buljac, Amine Bouterf, François Hild, and Stéphane Roux

# Chapter 1
# Digital Projection Speckle Technique for Fringe Generation

Austin Giordano, Andrew Nwuba, and Fu-Pen Chiang

**Abstract** Shadow moiré technique has been a widely utilized method in industry to determine surface flatness, out-of-plane displacement, and 3D metrology. In this paper, we present a digitized speckle method analogous to the shadow moiré method. A randomly generated speckle pattern is first projected onto a screen and digitally recorded. The same pattern is projected onto a specimen and digitally recorded as well. The two images are then converted into TIFF files to be superimposed and processed using a Fourier transform based algorithm to generate fringes that are similar to shadow moiré fringes. Additionally, a Digital Image Correlation (DIC) software was used to generate fringes that are similar to shadow moiré fringes. The technique is simple and straightforward. We apply the technique to a variety of specimens to demonstrate its applicability.

**Keywords** Optical metrology · Digital speckle photography · Digital image correlation · Moiré methods · Shadow moiré

## Introduction

The shadow moiré technique has been a ubiquitous technique for determination of out of plane displacements, both in terms of loading conditions and as a tool for metrological studying of the flatness of a surface [1]. The principles of white light speckle photography first described by Chiang and Asundi [2], later Chiang [3] showed the applicability of this technique and its integration with computational software. A new take on the white light speckle technique is presented where the speckles are projected onto the object of interest as opposed to being adhered to the surface. In this paper, a whole-field technique "out-of-plane" shape measurement is described in which the specimen's surface can be of any texture. With this method, no surface preparation is necessary, the only adjustment that would be recommended is a change in the color of the speckle pattern to achieve maximum contrast between the speckle pattern and the surface of the specimen.

## Experimental Procedures

The traditional shadow moiré technique was utilized to provide a baseline for the new proposed method. The optical arrangement for the shadow moiré technique is as shown in Fig. 1.1a. The system used for the data acquisition for the proposed digital projection speckle technique is as shown in Fig. 1.1b, c. The specimen is illuminated by a high-resolution projector, which is projecting a randomly generated speckle pattern. An image of the specimen with the speckle pattern projected onto it is captured by a digital camera. Another image is captured of just the speckle pattern being projected onto a projection screen. The white and black speckles are digitized into an array of 8688 × 5792 pixels. The image processing software ImageJ [4] is then used to convert the JPG image files into 8-bit TIFF files. The 8-bit TIFF files are then loaded into an image correlation software based on Fourier transforms [5] hereby referred to as CASI (Computer Aided Speckle Photography), and processed. Basic processes of the technique involve data acquisition and image processing. In the data acquisition stage, two speckle patterns, one with the specimen and one without the specimen, are captured by the camera and stored on an SD card, which is then registered into the computer. The image processing stage consists of four steps with ImageJ and CASI. First, the JPG image files are converted into 8-bit TIFF files which are then input into CASI. The three steps of the process that CASI is necessary for was described by Chen and Chiang [5]. First, an equivalent double-exposure

---

A. Giordano (✉) · A. Nwuba · F.-P. Chiang
Department of Mechanical Engineering, College of Engineering and Applied Sciences, State University of New York at Stony Brook, Stony Brook, NY, USA
e-mail: austin.giordano@stonybrook.edu; andrew.nwuba@stonybrook.edu; fu-pen.chiang@stonybrook.edu

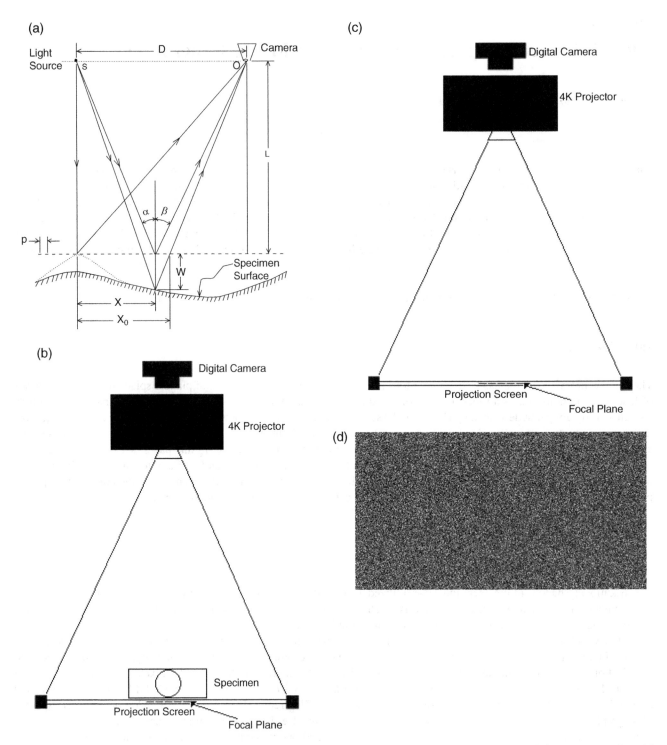

**Fig. 1.1** (a) Optical arrangement for the shadow moiré technique [1]. (b) Schematic of the data-acquisition system with the specimen in place. (c) Schematic of the data-acquisition system without the specimen. (d) Sample of speckle-pattern used

speckle pattern is obtained by superimposing the two digital images, and it is segmented into a series of small subimages. Second, a fast-Fourier transform (FFT) is applied to each subimage and "compared" with a correlation calculation. Last, a second FFT is applied to the resulting value giving rise to an impulse function whose position vector is nothing but the displacement vector experienced collectively by all the speckles contained within the subimage. Repeating the last two steps to all other subimages, we can resolve the full-field 2D displacements. CASI was utilized to provide quick feedback as to the validity of the experimental technique before more computational time was dedicated to using the DIC software Ncorr [6].

## Principle of the Method

The traditional shadow moiré fringes are formed by the shadow of the grating that is distorted to the surface of the specimen. The moiré fringes represent the topology of the specimen. From Fig. 1.1a the distance $w$ from the grating plane to the surface of the specimen for the shadow moiré can be expressed as:

$$w = \frac{Np}{\tan \alpha + \tan \beta} \quad (1.1)$$

where N is the fringe order; p is the grating pitch, $\alpha$ and $\beta$ are the illuminating and receiving angles from the normal to the grating plane. In the special case where both the light source and recording camera are at infinity and $\beta = 0$, Eq. (1.1) is reduced to:

$$w = \frac{Np}{\tan \alpha} \quad (1.2)$$

An analogous $w$ can be produced utilizing a double exposure approach to the proposed speckle technique. By importing two images, one of the speckle pattern projected onto the specimen and another of only the speckle pattern projected onto a projection screen into a DIC software the displacement of the speckles can be computed and a field analogous to the moiré fringes is produced.

## Results

In this experiment, we tested the proposed digital-speckle technique against the shadow moiré technique as a baseline for the results. The experiment was performed on a variety of different specimens to test the validity of the proposed technique. The specimens utilized are a sphere with a diameter of 76.2 mm, a cylinder with a diameter of 120.65 mm, and a mannequin head that has a distance from the back of the head to the tip of the nose of 177.8 mm. The moiré fringes, as well as the displacement fringes produced by CASI and Ncorr for each of the specimens, are shown in Figs. 1.2, 1.3, and 1.4. Each of the figures shows a good agreement between the proposed speckle technique and the traditional shadow moiré technique. The fringes produced by CASI were utilized to obtain feedback about the optical arrangement and speckle density before more computational time was dedicated to the Ncorr. The Ncorr results depict the displacement of the speckle pattern from the surface of the specimen to the projection screen, in the case of the circle and the cylinder this displacement is nearly identical to the diameter of the respective specimen. The largest displacement produced by the speckle patterns projected onto the mannequin head and the projection screen is within 0.5 mm of the actual distance from the tip of the nose to the back of the mannequin's head.

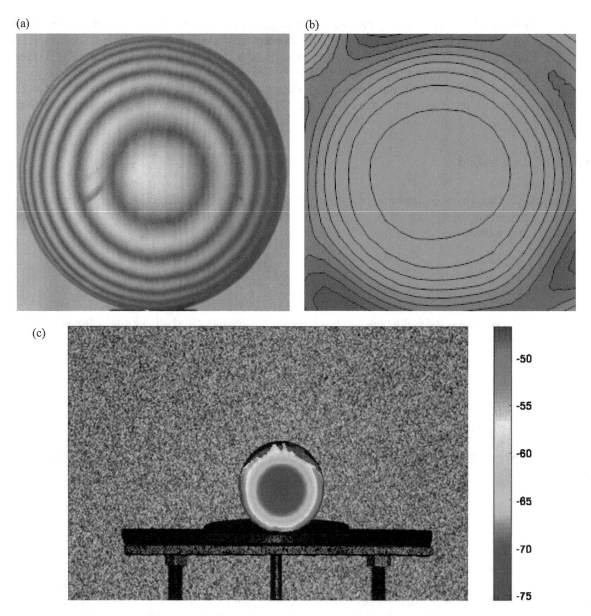

**Fig. 1.2** Fringe patterns of the topology of a sphere with diameter 76.2 mm (**a**) moiré fringes, (**b**) fringes produced with the speckle technique and CASI, (**c**) fringes produced with the speckle technique and Ncorr, legend in mm

## Conclusion and Discussion

A digital method has been proposed to use projected speckles to generate fringes that are analogous to shadow moiré fringes. By focusing the digital camera on a projection plane behind the specimen and capturing two images, one with the specimen in front of the projection plane, and one of only the projection plane. The resulting photographs are then Fast Fourier Transformed twice with the digital image correlation software CASI to produce fringes. Additionally, Ncorr the DIC software is utilized to produce additional fringes to prove the validity of the proposed technique.

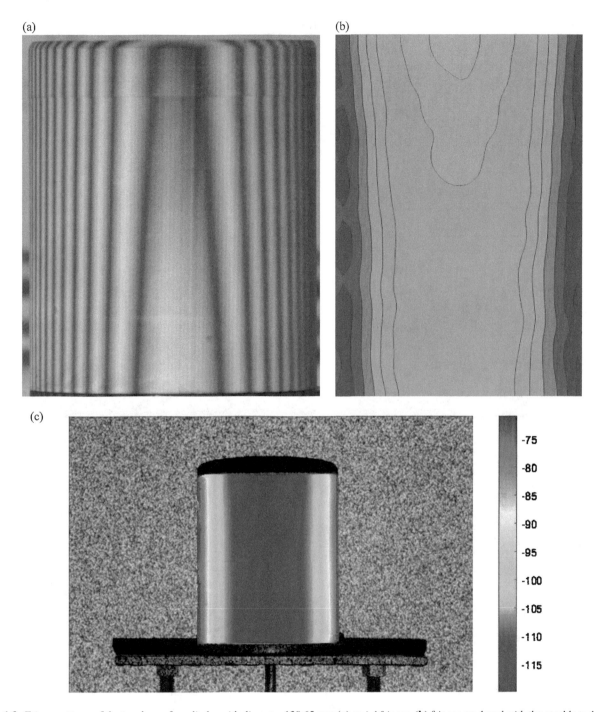

**Fig. 1.3** Fringe patterns of the topology of a cylinder with diameter 120.65 mm (**a**) moiré fringes, (**b**) fringes produced with the speckle technique and CASI, (**c**) fringes produced with the speckle technique and Ncorr, legend in mm

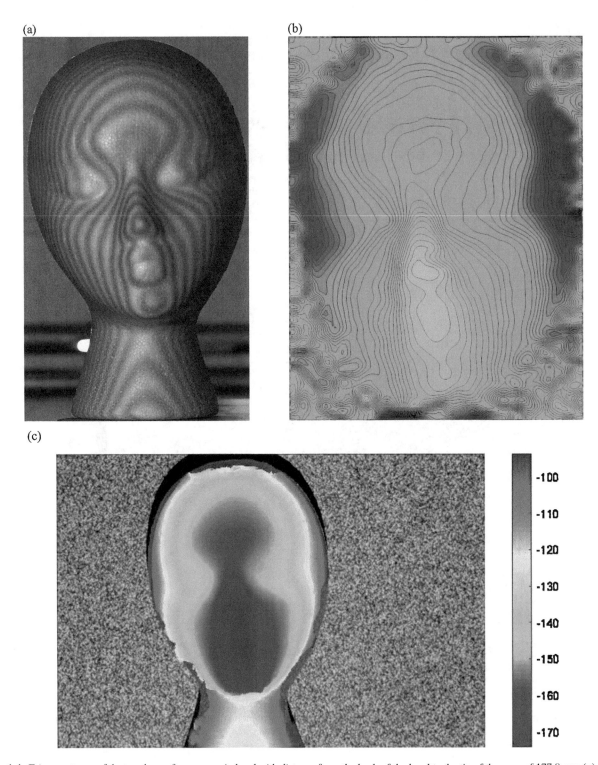

**Fig. 1.4** Fringe patterns of the topology of a mannequin head with distance from the back of the head to the tip of the nose of 177.8 mm (**a**) moiré fringes, (**b**) fringes produced with the speckle technique and CASI, (**c**) fringes produced with the speckle technique and Ncorr, legend in mm

**Acknowledgments** We gratefully acknowledge the support of Dr. Rajapakse, and the financial support of the Office of Naval Research through award no. -N00014-17-1-2873.

## References

1. Chiang, F. P. (1979). Chapter 6: Moiré methods of strain analysis. In *Manual on experimental stress analysis* (3rd ed., pp. 51–69). SESA.
2. Chiang, F. P., & Asundi, A. K. (1982). Objective white-light speckles and their application to stress-intensity factor determination. *Optics Letters, 7*, 378–379.
3. Chiang, F-P. (2003). Evolution of white light speckle method and its application to micro/nanotechnology and heart mechanics. *Optical Engineering 42*(5).
4. Schneider, C. A., Rasband, W. S., & Eliceiri, K. W. (2012). NIH image to ImageJ: 25 years of image analysis. *Nature Methods, 9*, 671–675.
5. Chen, D. J., & Chiang, F. P. (1993). Computer-aided speckle interferometry using spectral amplitude fringes. *Applied Optics, 32*, 225–236.
6. Blaber, J., Adair, B., & Antoniou, A. (2015). Ncorr: open-source 2D digital image correlation Matlab software. *Experimental Mechanics, 55*(6), 1105–1122.

# Chapter 2
# Quantifying Wrinkling During Tow Placement on Curvilinear Paths

**Sreehari Rajan, Michael A. Sutton, Roudy Wehbe, Brian Tatting, Zafer Gürdal, Addis Kidane, and Ramy Harik**

**Abstract** StereoDIC is employed to quantify wrinkling on 6.35 mm wide and 0.16 mm thick carbon-epoxy tows during advanced fiber placement along straight and circular paths on a planar composite surface. Measurements obtained just after placement provide quantitative deformation fields, including the presence of out-of-plane wrinkles that occur during the placement process. Results show that wrinkles occur at locations where the substrate has defects. Results also show that wrinkles occur at other locations only when the radius of curvature is less than 2540 mm.

**Keywords** StereoDIC · Automated fiber placement · Curvilinear paths · Wrinkling

## Introduction

The advent of advanced fiber placement systems allows composite manufacturers to smoothly vary the orientation of fibers in the component so that material properties can be optimally designed. Recent studies have shown that appropriate reorientation of fibers through placement along curvilinear paths in a composite [1, 2] can result in improved performance of the resulting laminate.

For composite material systems with high modulus fibers, tow steering has the potential to induce buckling of the fibers, separation of the tow from the underlying substrate and local tow uplift/wrinkling. This study presents experimental measurements using StereoDIC [3–7] to quantify the response of tows that are steered along curvilinear paths and bonded to the underlying substrate, with special emphasis on identifying local wrinkling during placement.

## Experiments

Stereovision with StereoDIC was employed in this work to extract deformation measurements from pairs of digital images. The parameters for the stereovision system hardware are given in Table 2.1. StereoDIC analysis of the images was performed using VIC-3D [8].

As shown in Table 2.1, lens distortion correction is performed with a third order radial distortion function. In addition, all analyses were performed with a $25 \times 25$ pixel$^2$ subset, spacing of 7 pixels between subsets, strain filter with $5 \times 5$ data points (area of $35 \times 35$ pixels$^2$) with central Gaussian weighting and the Lagrangian large strain tensor definition to determine all strain components.

To obtain high spatial resolution data across the width of the tow, both the camera magnification and applied speckle pattern size were adjusted. The optical field of view and speckle size for all experiments are given in Table 2.2. Since the field of view is limited to a length of ≈115 mm, and the manufacturing process lays up over 700 mm of tow, the stereovision system was translated along the length to acquire several, overlapping images of the tow before and after tow placement. To combine the individual images and present the results on a single combined image, image stitching is performed using the VIC-3D software.

Tow placement was performed using 6.35 mm wide carbon-epoxy prepreg slit tape. Each tow uses TORAYCA® T800S-24K carbon fibers impregnated with a thermoset binder. All placement was performed on a commercial Ingersoll AFP system in

---

S. Rajan · M. A. Sutton (✉) · R. Wehbe · B. Tatting · Z. Gürdal · A. Kidane · R. Harik
McNair Aerospace Center and Department of Mechanical Engineering, University of South Carolina, Columbia, SC, USA
e-mail: sreehari@email.sc.edu; sutton@sc.edu; rwehbe@email.sc.edu; tatting@cec.sc.edu; gurdal@cec.sc.edu; kidanea@cec.sc.edu; harik@mailbox.sc.edu

**Table 2.1** Camera and lens parameters for stereovision system

|  | AFP layup of tow |
|---|---|
| Camera | 5 MP CMOS PointGrey camera<br>• 2448 × 2048 pixels$^2$ array<br>• 3.45 μm pixel size |
| Calibration | 12 × 9 dot grid<br>• 5 mm dot size<br>• 1.5 mm dot spacing<br>More than 120 stereo image pairs |
| Lens | Nikon micro-Nikkor 25 mm focal length |
| Lens filter | Linear polarizer |
| Light source | LED with linear polarizing film |
| Lens distortion | 3rd order radial distortion correction |

**Table 2.2** Field of view and average speckle size

| Test type | Field of view | Average speckle size |
|---|---|---|
| AFP lay-up | 114 mm × 95 mm | 0.15 mm (∼3 pixels) |

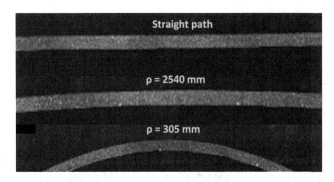

**Fig. 2.1** Photo of speckled tows and shapes of two circular paths

the McNair Aerospace center at the University of South Carolina. Process variables maintained throughout the layup process include (a) lay-down speed of 2438 mm/min, (b) tow heating to a temperature of 40° C and (c) compaction force of 1780 N. All the tows were placed on a substrate laminate of the same material [90/0/90]. Layup was performed for five paths and results are reported in this study for a straight path and for radii of curvature of 2540 and 305 mm, as shown in Fig. 2.1. Additional results are available in the literature for steering along curvilinear paths without adhesion [8] and with adhesion [9] to the substrate.

## Experimental Results and Discussion

Figures 2.2, 2.3, and 2.4 present the wrinkling/out-of-plane displacements for all three cases. Results shown in Figs. 2.2, 2.3, and 2.4 have several general features. First, tow wrinkles always occur where the tow crosses over a defect in the substrate (e.g., overlaps, gaps). Secondly, excluding the wrinkles occurring at substrate defects, for $\rho \geq 2540$ mm there were no significant wrinkles identified in the experiments. For $\rho \leq 1300$ mm, wrinkles were observed in the tow, with an increasing number of wrinkles occurring when $\rho = 305$ mm. Thirdly, excluding the uniformly placed wrinkles occurring at substrate defects, the spacing between wrinkles decreases with decreasing $\rho$. When the tow is steered with $\rho = 305$ mm, the average spacing is 21.30 mm ± 12.93 mm,

Observations made during tow placement clearly showed that wrinkling in the tow occurred just after the moving roller passed over the heated tow. Based on these observations, it is conjectured that viscoelastic effects may not contribute significantly to the response of the epoxy matrix material.

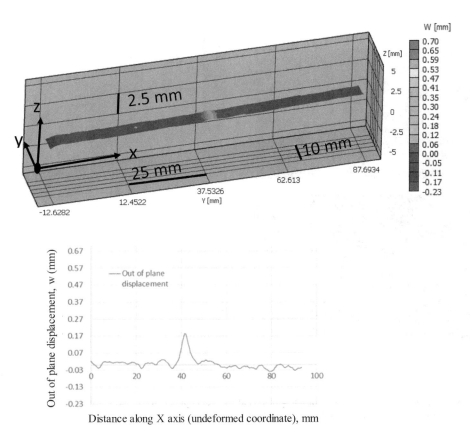

**Fig. 2.2** Out of plane displacements for one section of a tow placed along a straight path. Visible wrinkle is due to an overlap defect in the underlying substrate

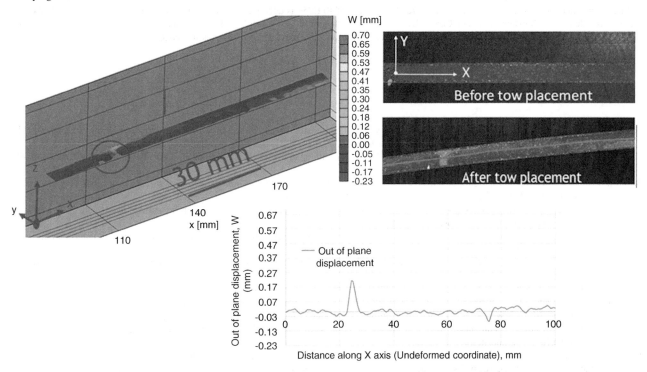

**Fig. 2.3** Out of plane displacements for one section of a tow placed along a circular arc with $\rho = 2540$ mm. Visible wrinkle is due to an overlap defect in the underlying substrate

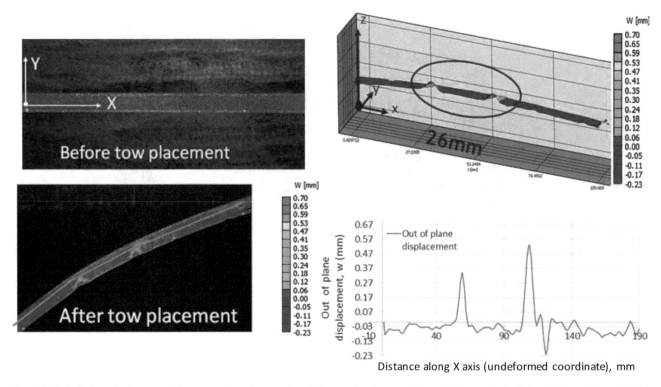

**Fig. 2.4** Out of plane displacements for one section of a tow placed along a circular arc with $\rho = 305$ mm. Both of the visible positive wrinkles are due to adhesive separation and are not at substrate defect locations

**Acknowledgements** Funding provided by Boeing Research Contract SSOWBRTW0915000 and associated matching funds provided by University of South Carolina Vice President for Finance Edward Walton via 15540 E250 is deeply appreciated. All materials and access to the Lynx® AFP facility provided by the McNair Aerospace Center, University of South Carolina is gratefully acknowledged. The technical support and assistance of the McNair technical staff, particularly Mr. Burton Rhodes, Jr. for tow placement, is gratefully acknowledged.

# References

1. Gürdal, Z., Tatting, B. F., & Wu, C. K. (2008). Variable stiffness composite panels: Effects of stiffness variation on the in-plane and buckling response. *Composites. Part A, Applied Science and Manufacturing, 39*, 911–922. https://doi.org/10.1016/j.compositesa.2007.11.015.
2. Gürdal, Z., & Olmedo, R. (1993). In-plane response of laminates with spatially varying fiber orientations: Variable stiffness concept. *AIAA Journal, 31*, 751–758. https://doi.org/10.2514/3.11613.
3. Sutton, M. A., McNeill, S. R., Helm, J. D., & Chao, Y. J. (2000). *Advances in two-dimensional and three-dimensional computer vision, photomechanics* (pp. 323–372). Berlin: Springer. https://doi.org/10.1007/3-540-48800-6_10.
4. Helm, J. D., McNeill, S. R., & Sutton, M. A. (1996). Improved three-dimensional image correlation for surface displacement measurement. *Optical Engineering, 35*, 1911–1920. https://doi.org/10.1117/1.600624.
5. Sutton, M. A. (2013). Computer vision-based, noncontacting deformation measurements in mechanics: A generational transformation. *Applied Mechanics Reviews, 65*, 050000. https://doi.org/10.1115/1.4024984.
6. Luo, P. F., Chao, Y. J., & Sutton, M. A. (1994). Application of stereo vision to three-dimensional deformation analyses in fracture experiments. *Optical Engineering, 33*, 3.
7. Luo, P. F., Chao, Y. J., Sutton, M. A., & Peters, W. H. (1993). Accurate measurement of three-dimensional deformations in deformable and rigid bodies using computer vision. *Experimental Mechanics, 33*, 123–132. https://doi.org/10.1007/BF02322488.
8. Rajan, S., Sutton, M. A., Wehbe, R., Tatting, B., Gürdal, Z., Kidane, A., & Harik, R. (2019). Experimental investigation of prepreg slit tape wrinkling during automated fiber placement process using StereoDIC. *Composites Part B: Engineering, 160*, 546–557. https://doi.org/10.1016/j.compositesb.2018.12.017.
9. Rajan, S., Sutton, M. A., Gurdal, Z., Tatting, B., Wehbe, R., & Kidane, A. (2019). Measured surface deformation and strains in thin thermoplastic prepreg tapes steered along curved paths without adhesion using StereoDIC. *Experimental Mechanics, 59*(4), 531–547.

# Chapter 3
# Experimental Mechanics, Tool to Verify Continuum Mechanics Predictions

**C. A. Sciammarella, L. Lamberti, and F. M. Sciammarella**

**Abstract** This paper is devoted to the experimental verification of a very fundamental concept in the mechanics of materials, the representative volume element (RVE). This concept is a bridge between the theoretical concept of the continuum and the actual discontinuous structure of matter. We begin with reviewing the pertinent concepts of the kinematics of the continuum, the mathematical functions that relate displacement vectorial fields, the recording of these fields by a sensor as scalar fields of gray levels.

The derivative field tensors corresponding to the Eulerian description are then connected to the deformation of the continuum. The differential geometry that provides the deformation of an element of area is introduced. From this differential geometry of an element of area, the Euler-Almansi tensor is extracted. Properties of the Euler-Almansi tensor are derived. The next step is the analysis of the relationship between kinematic and dynamic variables: that is, the connection between strains and stresses in the Eulerian description between the Euler-Almansi tensor with the Cauchy stress tensor.

In the experimental part of the paper, some relationships between components of the Euler-Almansi tensor are verified. An example of an experimental verification of the concept of RVE is given. Finally, the verification for the fact that the Euler-Almansi and Cauchy stress sensor tensors are conjugate tensors in the Hill-Mandel sense is presented.

**Keywords** Representative volume element (RVE) · Statistical volume element (SVE) · Kinematical variables · Derivatives of displacements · Euler-Almansi strain tensor · Cauchy stress tensor

## Introduction

A very important aspect of Experimental Mechanics both from the point of view of practical engineering applications as well from the theoretical developments in the mechanics of materials is to establish the validity of the predictions of Continuum Mechanics in view of the heterogeneous structure of matter. The continuum hypothesis provides answers on the behavior of materials that are based on the concept of representative volume element (RVE) or statistical volume element (SVE). A representative RVE implies a bridge between probabilistic outcomes based on the theory of stochastic variables and a deterministic outcome based on physical laws. In a simplified scheme, we can explain the idea of RVE in the following way. One determines the strain $\varepsilon$, for example with strain gages. On a certain volume of a solid medium, one obtains an outcome, a given value of $\varepsilon$. This value is tied to a volume of the material, which in turn is related to a given size L. This volume is composed of micro-components with a size characterized by a length d. The ratio $\delta = L/d$ is called in the literature mesoscale. For a given $\delta$, one gets different responses $\varepsilon$ for different materials because of the different "d" for each material. For a given L as $\delta$ changes, the responses of the different volumes converge to a certain value $\varepsilon_{rve}$, Fig. 3.1, the subscript "rve" indicating RVE and the scatter of the different values disappears and gives a deterministic value, within a certain number of significant figures. In the literature, this behavior is called ergodicity and corresponds to the statistical analysis of random variables.

---

C. A. Sciammarella (✉)
Department of Mechanical, Materials and Aerospace Engineering, Illinois Institute of Technology, Chicago, IL, USA

Department of Mechanical Engineering, Northern Illinois University, DeKalb, IL, USA
e-mail: sciammarella@iit.edu

L. Lamberti
Dipartimento Meccanica, Matematica e Management, Politecnico di Bari, Bari, Italy
e-mail: luciano.lamberti@poliba.it

F. M. Sciammarella
Department of Mechanical Engineering, Northern Illinois University, DeKalb, IL, USA
e-mail: sciammarella@niu.edu

**Fig. 3.1** Local strain value as a function of the representative volume scale for a given δ value

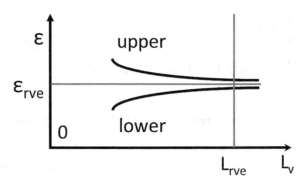

The example we are analyzing deals with the values of $\varepsilon_{rve}$ that can be measured by using different sizes of strain gages and to the fact that these values can converge to a single value. In this case, ergodicity implies that for processes that are not time dependent, the spatial average, $\varepsilon_{rve}$ is equal to the ensemble average. This means that if one measures strains with a scale of order of magnitude of the RVE, the average of these strains will converge to a single value of $\varepsilon_{rve}$ and of course this convergence must occur at any RVE of the analyzed material and this property defines the homogeneity of the material. There are upper limits and lower limits to the RVE. The RVE should be of a size capable of capturing what is called local value of the considered variable. For example, in the case of strain $\varepsilon$, the strain variation in the considered volume must be stationary, that is, it varies almost like a small slope linear function in the volume under consideration. This means that the local values have negligible changes if the representative volume is changed from some optimum value. This idea has a simple graphical representation in Fig. 3.1.

One can start from a lower bound or from an upper bound. As the characteristic length reaches the $L_{rve}$ value, the $\varepsilon_{rve}$ reaches an asymptotic value and it is possible to select a number of significant figures that define the local value. The above theoretical considerations are very important and should be considered when one is dealing with optical techniques where the analyzed volumes are spatially sampled. For techniques that utilize carrier signals, the sampled volumes will depend on the pitch p of the tagged signals, and on the added carrier required to make the displacement determination feasible. In the case of DIC, the representative volume will depend on the sub-elements size and the number of pixels on the sub-elements. These considerations apply to kinematic properties like strains but also to other properties where the concept of force is involved, the stress $\sigma$. The required conditions are given on the basis of the Hill-Mandel homogenization condition. It involves the space of admissible displacements in the RVE. Speaking in general terms, it implies to say that for $\sigma_{rve}$ and $\varepsilon_{rve}$ the virtual work in the macroscale equals the virtual work in the subscale. This condition must be considered when constitutive functions are derived to get a match between selected kinematic variables and corresponding dynamic variables that are compatible with each other.

With reference to notation, we consider real or complex-valued functions f(**x**) defined on $\Re_n$, where n = 1,2. Ordinary case letters represent scalar quantities, bold letters represent vectorial quantities. It will be written f(**x**) or f(x,y), where the bold lower case indicates a vector quantity or we will list low-case variables, whichever is more convenient in the context of the discussion.

## Determination of the Local Kinematic Variables

Let us begin with an overview of the local kinematics of the 2D continuum for large deformations and large rotations. There are very important considerations that must be made when proceeding to obtain local values of displacements and displacement derivatives. These considerations are connected with the selection of the strain tensor that is valid for a given RVE. In our initial studies dealing with fringes corresponding to projected displacements (moiré fringes or isothetic lines), the linearized strain tensor was the basis for the recovery of displacements and displacement derivatives. In [1] and in the references contained in this paper, the OSA method for fringe pattern analysis is described. In [1], the concept of RVE is indirectly introduced by defining all the kinematic variables for local values which are determined by the selection of the scale L, the pitch of the tagged carrier p, and the introduced carrier fringes originated by adding a carrier using frequency shifting or other procedures.

Figure 3.2a represents the local region (RVE) of the vertical displacements, and the RVE satisfies the condition shown graphically in Fig. 3.1. The displacement vector **V** that should be vertical due to presence of rigid body rotation has an important horizontal component $v_x$. The value of the angle $\alpha_v$ that gives the direction of the fringes should be $\alpha_v = 0$ if

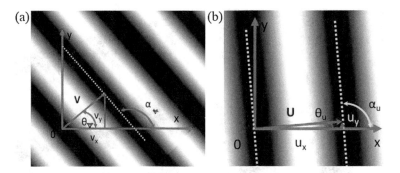

**Fig. 3.2** (**a**) Local pattern of vertical displacement V (horizontal fringes in the undeformed condition) for large deformations and rotations. (**b**) Small deformations and rotations for horizontal displacement U

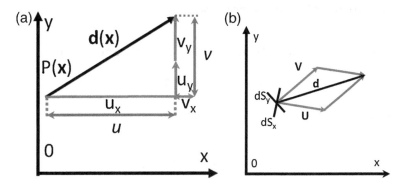

**Fig. 3.3** (**a**) Vector displacement **d(x)** and its projected components; (**b**) Vectorial sum of the two component vectors

there is no rotation component, this value measured counterclockwise with respect to the x-axis, in Fig. 3.2a is $\alpha_v > \pi/2$. In Fig. 3.2b corresponding to the **U** displacements, $\alpha_u$ should be $\alpha_u = \pi/2$, the actual angle is slightly bigger than $\pi/2$, $\theta_u \approx 0$, and the modulus of **U** is approximately equal to $u_x \approx u$ the u indicating the projection of the horizontal displacement in the x-axis, $u_y$ can be neglected. Let us consider now the displacement vector **d(x)** at a given point P(**x**), see Fig. 3.3.

For large deformations and large rotations, we have the relationship [1],

$$\begin{cases} d_x(x) = u_x(x) + v_x(x) = u(x) \\ d_y(x) = u_y(x) + v_y(x) = v(x) \end{cases} \quad (3.1)$$

The above equations are the generalized relationships for large deformations and rotations corresponding to the equations utilized for moiré fringes or isothetic lines.

## Transformation of Recorded Gray Levels into Vectorial Fields

Let us consider [1] the process of transformation of recorded images, that are scalar functions of gray levels, into vectorial field corresponding to displacements. The recorded image consists of a scalar potential of gray levels $F_{2R}$, the subscript 2 indicates a 2D function and R indicates that is a real quantity. Introducing carrier fringes and adopting Cartesian coordinates, the potential scalar functions has two scalar components,

$$F_{2R}(\mathbf{x}) = U(\mathbf{x}) + V(\mathbf{x}) \quad (3.2)$$

The two scalar functions are carrier fringes that through the applications of the Fourier transform or the Hilbert transform [1] are converted into two vectorial functions, Fig. 3.3b,

$$\mathbf{d}(\mathbf{x}) = \mathbf{U}(\mathbf{x}) + \mathbf{V}(\mathbf{x}) \quad (3.3)$$

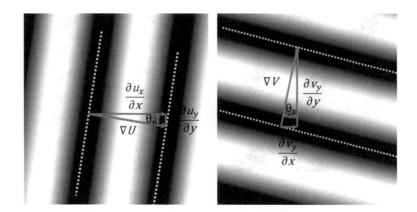

**Fig. 3.4** Local gradients vectors and corresponding Cartesian components

If one applies the gradient operation to Eq. (3.2) and utilizes the distributive property of the gradient,

$$\nabla F_{2R} = \nabla U(x) + \nabla V(x) \tag{3.4}$$

That results in the projection equations,

$$\begin{cases} \frac{\partial d_x(x)}{\partial x} = \frac{\partial u_x(x)}{\partial x} + \frac{\partial v_x(x)}{\partial x} = \frac{\partial u}{\partial x} \\ \frac{\partial d_y(x)}{\partial y} = \frac{\partial u_y(x)}{\partial y} + \frac{\partial v_y(x)}{\partial y} = \frac{\partial v}{\partial x} \end{cases} \tag{3.5}$$

The components of Eq. (3.5) are illustrated in Fig. 3.4. It can be seen that the vectors $U(x)$ and $V(x)$ are co-axial with the vectors $\nabla U(x)$ and $\nabla V(x)$, only the scales are changed.

Returning to Eq. (3.4), it also can be written,

$$\nabla F_{2R} = \frac{\partial F_{2R}}{\partial x} i + \frac{\partial F_{2R}}{\partial y} j \tag{3.6}$$

Equation (3.6) represents a vector, the gradient vector, let us call it $G_2(x)$. Computing the divergence of this vector, one gets,

$$\nabla \cdot G_2(x) = \frac{\partial^2 F_{2R}}{\partial^2 x} + \frac{\partial^2 F_{2R}}{\partial^2 y} \tag{3.7}$$

There is another important function of a vectorial field. The rotor or curl of the field that is defined by,

$$\nabla \times G_2 = \left( \frac{\partial^2 F_{2R}(x)}{\partial y \partial x} - \frac{\partial^2 F_{2R}(x)}{\partial x \partial y} \right) k \tag{3.8}$$

If the field of displacements is such that there are no rotations, it follows

$$\nabla \times G_2 = \left( \frac{\partial^2 F_{2R}(x)}{\partial y \partial x} - \frac{\partial^2 F_{2R}(x)}{\partial x \partial y} \right) k = 0 \tag{3.9}$$

In the case that the divergence and the curl of the fields are zero, then

$$\begin{cases} \frac{\partial^2 F_{2R}}{\partial^2 x} = -\frac{\partial^2 F_{2R}}{\partial^2 y} \\ \frac{\partial^2 F_{2R}}{\partial y \partial x} = \frac{\partial^2 F_{2R}}{\partial x \partial y} \end{cases} \tag{3.10}$$

From Eq. (3.2), one obtains,

$$\begin{cases} \frac{\partial^2(U(x)+V(x))}{\partial^2 x} = -\frac{\partial^2(U(x)+V(x))}{\partial^2 y} \\ \frac{\partial^2(U(x)+V(x))}{\partial y \partial x} = \frac{\partial^2(U(x)+V(x))}{\partial x \partial y} \end{cases} \tag{3.11}$$

From Eqs. (3.10) and (3.11) it follows that the gray level scalar potential $F_{2R}(x)$ that defines the local phase must be a solution of the Laplace's equation,

$$\frac{\partial^2 F_{2R}(x)}{\partial^2 x} + \frac{\partial^2 F_{2R}(x)}{\partial^2 y} = 0 \tag{3.12}$$

The solutions of the Laplace equation are part of the theory of potentials. These solutions are known to be harmonic functions. To give a correct interpretation to Eq. (3.12), it is necessary to remember that in the developments presented in this paper we are dealing with the concept of local values corresponding to the RVE, including the fact that in order to utilize the adopted model of the light signal it is necessary to have carrier fringes of frequency higher than the local values of the signal [1]. These conclusions are graphically represented in Figs. 3.2 and 3.4 where the local values of the gray levels are harmonic functions.

When displacement fields do not satisfy the conditions of being irrotational and conservative, then the solutions of the 2D continuum is given by the Poisson's equation,

$$\frac{\partial^2 F_{2R}(x)}{\partial^2 x} + \frac{\partial^2 F_{2R}(x)}{\partial^2 y} = P_s(x) \tag{3.13}$$

In Eq. (3.13), $P_s(x)$ is called a source term. These two partial differential equations provide solutions for different types of displacement fields in 2D.

The above derivations agree with the Continuum Mechanics analytic solutions of 2D problems and provide tools for the experimental determination of the displacement fields and their derivatives.

## Analysis of the Displacement Functions in the Complex Plane

Having derived from the point of view of recorded images as scalar gray levels the mathematical relationships between scalar potential functions and displacement vectors, it is necessary to connect these potentials with the tools utilized by the OSA method to get the actual components of the displacement vector from the recorded gray levels. To achieve this objective, we need to move from the physical space to the complex space.

The scalar gray levels in the physical space are associated with complex potential in the complex plane [1]. This relationship between the two planes brings some interesting results concerning methods utilized to retrieve displacements from scalar fields of gray levels.

Let us consider complex plane $\mathcal{C}$ and introduce the complex potentials $U_C^e(x_c)$ and $V_C^o(x_c)$. The upper script "e" means that the potential is an even function and "o" means that the potential is and odd function. One defines two complex potentials,

$$\begin{cases} z_{xc}(x_c) = U_c^e(x_c)\,\vec{i} + U_c^0(x_c)\,\vec{j} \\ z_{yc}(x_c) = V_c^e(x_c)\,\vec{i} + V_c^o(x_c)\,\vec{j} \end{cases} \tag{3.14}$$

In Eq. (3.14), the symbol $\Rightarrow$ is utilized to indicate that the versors correspond to the complex plane. Assuming that each of these components are real and have derivatives at a given point in the complex plane, it is known that each of these functions must satisfy in the complex plane the Cauchy-Riemann conditions,

$$\begin{cases} \frac{\partial U_c^e}{\partial x_c} = \frac{\partial U_c^o}{\partial y_c} \\ \frac{\partial U_c^e}{\partial y_c} = -\frac{\partial U_c^o}{\partial x_c} \end{cases} \tag{3.15}$$

Similar equations can be written for the other potential function corresponding to the V components. Through Eq. (3.15), one gets the mathematical connection between the Hilbert transform, holomorphic functions and the levels of gray as potential functions that lead to the definition of a local phase. Thus, a consistent mathematical framework is adopted that supports the OSA methodology of displacement retrieval utilizing gray levels recorded in an image.

For example, if $U_c^o(x)$ is of the form

$$U_{xc}^e(x_c) = I_p \cos \phi (x) \qquad (3.16)$$

Applying the Hilbert transform one obtains

$$U_{xc}^o(x_c) = I_q \sin \phi (x) \qquad (3.17)$$

We know [1] that these relationships are approximately valid in displacement fields and it is required that in the REV these functions are approximately stationary, as shown in Fig. 3.1. The above derivations are graphically represented by the Poincare sphere [1], a 3D plot in the complex plane that illustrates the necessary components needed to experimentally determine the state of deformation of a pixel in the 2D physical space.

## Derivatives of the Displacements

From the displacement field we now move to the derivatives of the displacements that form an essential quantity in the solution of the Experimental Mechanics problems. The derivatives [1] can be obtained directly from gray levels without previously retrieving the displacement vectorial field. In [2], there is a review of the fundamental concepts of Continuum Mechanics concerning the description of the deformation of a medium and the derivatives of the displacements. In this section, we will analyze the Eulerian description and briefly look over the mathematics relevant to the derivations that will be introduced later. One condition that the derivatives of the displacements must satisfy is that the Jacobian of the coordinate change in 2D is different from zero [2] and we will continue with the analysis of 2D fields,

$$J_2 = \det \begin{vmatrix} \frac{\partial x}{\partial X} & \frac{\partial x}{\partial Y} \\ \frac{\partial y}{\partial X} & \frac{\partial y}{\partial Y} \end{vmatrix} \neq 0 \qquad (3.18)$$

In Eq. (3.18), (x,y) are the Eulerian coordinates and X,Y are the Lagrangian coordinates of a point of the continuum. The displacement vector of a point has two components: one is the relative change of distance between the point and a neighbor point, the other is a rigid body displacement due to rigid body translation and rotation caused by the deformation of the rest of the body where the point under analysis is located. The vector displacement in 2D is defined according to the notation introduced in Eq. (3.1),

$$\begin{cases} d_x = u \\ d_y = v \end{cases} \qquad (3.19)$$

In the final step of the local transformation one arrives to a tensor of the derivatives [2],

$$[J] = \begin{bmatrix} \frac{\partial u}{\partial x} & \frac{\partial u}{\partial y} \\ \frac{\partial v}{\partial x} & \frac{\partial v}{\partial y} \end{bmatrix} \qquad (3.20)$$

From this tensor one can obtain all the necessary information to describe the deformation in the Eulerian description in 2D. The tensor Eq. (3.20) can be split into a symmetric tensor and an anti-symmetric tensor. The former is,

$$[J_s] = \begin{bmatrix} \frac{\partial u}{\partial x} & \frac{1}{2}\left(\frac{\partial u}{\partial y} + \frac{\partial v}{\partial x}\right) \\ \frac{1}{2}\left(\frac{\partial u}{\partial y} + \frac{\partial v}{\partial x}\right) & \frac{\partial v}{\partial y} \end{bmatrix} \qquad (3.21)$$

The anti-symmetric tensor is,

$$[J_a] = \begin{bmatrix} 0 & \frac{1}{2}\left(\frac{\partial u}{\partial y} - \frac{\partial v}{\partial x}\right) \\ \frac{1}{2}\left(\frac{\partial u}{\partial y} - \frac{\partial v}{\partial x}\right) & 0 \end{bmatrix} \quad (3.22)$$

The tensor $J_s$ contains the derivatives related to the deformation components of the continuum, and tensor $J_a$ provides the derivatives related to the rigid rotation of the considered element of surface.

## Displacement Vector and Metric Properties

To visualize the local transformations that take place due to the applied deformation to a continuum medium is useful a graphical representation that applies to the Eulerian description, Fig. 3.5. We have a Cartesian reference system that represents both the Lagrangian coordinate system and the Eulerian coordinate system. Both systems have parallel versors to the Cartesian reference system. A point $M_0$ moves to the point $M_1$, and the vector joining these two positions of the point, is the vector $d(x)$, the same vector in both coordinate systems [2]. The arc elements that upon deformation become parallel to the Eulerian coordinate system are given by two vectors $U_L(x) = \overline{M_0 N_0}$ and $V_L(x) = \overline{M_0 P_0}$.

Applying differential geometry to the changes experienced by the medium and symbolically represented in Fig. 3.5, one obtains the Eulerian Almansi tensor that fulfils the condition of given zero components upon a rigid body rotation.

$$\begin{cases} \varepsilon_x^E = 1 - \sqrt{1 - 2\frac{\partial u}{\partial x} + \left(\frac{\partial u}{\partial x}\right)^2 + \left(\frac{\partial v}{\partial x}\right)^2} \\ \varepsilon_y^E = 1 - \sqrt{1 - 2\frac{\partial v}{\partial y} + \left(\frac{\partial v}{\partial y}\right)^2 + \left(\frac{\partial u}{\partial y}\right)^2} \\ \left(\varepsilon_{xy}^E\right) = \arcsin \frac{\frac{\partial u}{\partial y} + \frac{\partial v}{\partial x} - \frac{\partial u}{\partial x}\frac{\partial u}{\partial y} - \frac{\partial v}{\partial x}\frac{\partial v}{\partial y}}{(1 - \varepsilon_x^E)(1 - \varepsilon_y^E)} \end{cases} \quad (3.23)$$

It is interesting to point out that the experimental determinations are done in the deformed geometry. If one is dealing with the linear tensor, the assumption is that $\theta_{xx}$ and $\theta_{yy}$ are negligible quantities. It is also assumed that the derivatives of the displacements are quantities much smaller than 1, hence their squares and products are of an order of magnitude such that can be neglected. Thus, the difference between the initial and final geometries can be disregarded. One can operate with

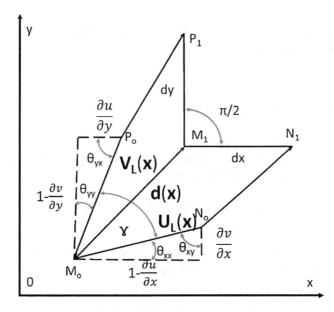

**Fig. 3.5** Differential geometry representation of the effect of the changes of the projections of the displacement vector in the local deformation at a point of the continuum medium

the geometry of the initial configuration. This is no longer the case if the previously mentioned quantities do not satisfy the specified conditions. In this case, the Lagrangian and the Eulerian descriptions lead to different results. The Eulerian description is the most adequate approach in Experimental Mechanics because reflects the physical reality of the actual measurements.

There is another version of the tensor, sometimes in the literature referred also as the Euler-Almansi tensor,

$$\begin{cases} \varepsilon_x^E = \frac{\partial u}{\partial x} - \frac{1}{2}\left[\left(\frac{\partial u}{\partial x}\right)^2 + \left(\frac{\partial v}{\partial x}\right)^2\right] \\ \varepsilon_y^E = \frac{\partial v}{\partial y} - \frac{1}{2}\left[\left(\frac{\partial u}{\partial y}\right)^2 + \left(\frac{\partial v}{\partial y}\right)^2\right] \\ \varepsilon_{xy}^E = \frac{1}{2}\left[\frac{\partial u}{\partial y} + \frac{\partial v}{\partial x} - \frac{\partial u}{\partial x}\frac{\partial u}{\partial y} - \frac{\partial v}{\partial x}\frac{\partial v}{\partial y}\right] \end{cases} \quad (3.24)$$

This tensor does not become zero upon a rigid body rotation. It is possible to show, for example, that upon a rotation $\Theta$, it occurs,

$$\varepsilon_x^E = \frac{\sin^2\theta}{2} \quad (3.25)$$

This expression is not zero, but if $\theta$ is small then $\varepsilon_x^E \approx 0$. Hence, the Eq. (3.24) are approximations that are valid in a range of small rotations that are not small enough to justify the linear tensor but not large enough that is possible to neglect the remainder error caused by the simplification of the tensor given in Eq. (3.24).

## Rigid Body Rotation of an Element of Area

We have dealt with the symmetric part of the tensor of the derivatives, Eq. (3.21), that handles the deformation of the medium. We need to consider the anti-symmetric part that is related to a relative local rigid body rotation of the considered element of volume. In Continuum Mechanics, the local rigid body rotation of an element of volume $\Omega(\mathbf{x})$, is equal to one half of the curl of the local displacement field. Applying the definition of curl in 3D,

$$\Omega = \frac{1}{2}\begin{vmatrix} \mathbf{i} & \mathbf{j} & \mathbf{k} \\ \frac{\partial}{\partial x} & \frac{\partial}{\partial y} & 0 \\ u & v & 0 \end{vmatrix} = \text{arctg}\frac{1}{2}\left(\frac{\partial v}{\partial x} - \frac{\partial u}{\partial y}\right)\mathbf{k} \quad (3.26)$$

The difference of the cross-derivatives corresponds to the rigid body rotation of an element of area or from the point of view of the 3D space, the rotation of an element of volume.

## Properties of the Symmetric Part of the Tensor

It is interesting to analyze further the properties of the symmetric tensor of Eq. (3.21). The sum of principal derivatives is an invariant of the tensor. If one considers now the principal directions of the derivatives of the tensor for a system of coordinates aligned with the principal directions and calling the corresponding vectors $\mathbf{S}_1$ and $\mathbf{S}_2$, one can write,

$$\frac{\partial u}{\partial x} + \frac{\partial v}{\partial y} = \frac{\partial u}{\partial s_1} + \frac{\partial v}{\partial s_2} = \varepsilon_{p1}^E + \varepsilon_{p2}^E \quad (3.27)$$

In Eq. (3.27), $s_1$ and $s_2$ are the coordinates measured along the corresponding principal directions.

The normal strain in the x-direction is given in the tensor Eq. (3.23),

$$\varepsilon_x^E = 1 - \sqrt{\left(1 - \frac{\partial u}{\partial x}\right)^2 + \left(\frac{\partial v}{\partial x}\right)^2} \quad (3.28)$$

**Fig. 3.6** Relationship between the principal directions at a given point of the field, the gradient of the displacement vector, and the vector displacement **d(x)**

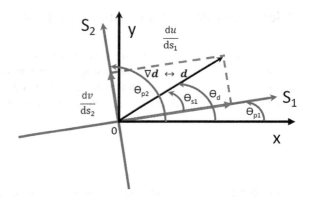

If the coordinate axes are the direction of the principal strains, and these directions are $\mathbf{S}_1$, $\mathbf{S}_2$ then, $\frac{\partial u}{\partial x} = \frac{\partial u}{\partial s_1}$ and $\frac{\partial v}{\partial x} = \frac{\partial v}{\partial s_1} = 0$, since the shear is zero for the principal strains. Then, Eq. (3.28) becomes,

$$\varepsilon_{p1}^E = 1 - \sqrt{\left(1 - \frac{\partial u}{\partial s_1}\right)^2} = \frac{\partial u}{\partial s_1} \tag{3.29}$$

A similar derivation can be made for the other principal strain. Then,

$$\begin{cases} \varepsilon_{p1} = \frac{du}{ds_1}\mathbf{S}_1 \\ \varepsilon_{p2} = \frac{dv}{ds_2}\mathbf{S}_2 \end{cases} \tag{3.30}$$

where $\mathbf{S}_1$ and $\mathbf{S}_2$ are the versors corresponding to the principal directions.

Equation (3.30) indicate that the derivatives of the projected displacements are the projection of the gradient of the vector **d(x)** that is co-axial with the vector. Then, from Eq. (3.5), we get the vector equation (Fig. 3.6),

$$\nabla d = \frac{du}{ds_1}\mathbf{S}_1 + \frac{dv}{ds_2}\mathbf{S}_2 \tag{3.31}$$

Summarizing previously presented developments and recalling Eq. (3.30), the sum of the principal strains is equal to the invariant $\frac{\partial u}{\partial s_1} + \frac{\partial v}{\partial s_2}$. The principal strains are computed using the equations,

$$\begin{cases} \varepsilon_{p1,2}^E = \frac{\varepsilon_x^E + \varepsilon_y^E}{2} \pm \sqrt{\left(\frac{\varepsilon_x^E - \varepsilon_y^E}{2}\right)^2 + \left(\frac{\gamma_{xy}^E}{2}\right)^2} \\ \text{tg } 2\theta_p = \frac{2\varepsilon_{xy}^E}{\varepsilon_x^E - \varepsilon_y^E} \end{cases} \tag{3.32}$$

The position of the displacement vector with respect to the principal direction $\mathbf{S}_1$ is given by,

$$\theta_{s1} = \text{arctg}\frac{\frac{dv}{dS_2}}{\frac{du}{dS_1}} = \frac{\varepsilon_{p2}^E}{\varepsilon_{p1}^E} \tag{3.33}$$

The following relationship between the direction of the vector **d(x)** and the direction of the principal strains can be obtained:

$$\theta_d = \theta_{p1} + \theta_{s1} \tag{3.34}$$

## Additional Information From the Kinematics of the Continuum

In previous sections, we have connected theoretical aspects of the continuum kinematics and the experimental observations for large rotations and deformations using two versions of the Euler-Almansi tensor. Now we are going to analyze additional relations that apply as the magnitude of rotations and deformations are reduced.

Going back to Eq. (3.23), if the rotations are small as well as the deformations, it follows, Fig. 3.5:

$$\begin{cases} \theta_{xx} = \frac{\partial v}{\partial x} \\ \theta_{yy} = \frac{\partial u}{\partial y} \end{cases} \quad (3.35)$$

Furthermore, if $\frac{\partial u}{\partial x}, \frac{\partial u}{\partial y} \ll 1$ and $\frac{\partial v}{\partial x}, \frac{\partial v}{\partial y} \ll 1$, their squares and products can be neglected. Then, Eq. (3.21), the symmetric part of the tensor of the derivatives $[J_s]$ becomes the strain tensor. Going back to the curl of the field, Eq. (3.26), the anti-symmetric part of the tensor of the derivatives that defines the local rigid body rotation is such that $\Omega \approx 0$ then,

$$\frac{\partial u}{\partial y} = \frac{\partial v}{\partial x} \quad (3.36)$$

The meaning of Eq. (3.36) is that the displacement field is irrotational. Furthermore, if there are no sources of displacements in the field, the divergence of the field is zero, hence the displacement field for small deformations and rotations satisfies the Cauchy-Riemann conditions, therefore is a solution of the Laplace equation, Eq. (3.12).

The invariance of the sum of the derivatives, Eq. (3.27), in the case of a specimen under plane stress allows to generalize relationship derived in [1] for the principal strains of the tensor,

$$\sigma_1 + \sigma_2 = \frac{E}{1-\nu}\left(\varepsilon_{p1} + \varepsilon_{p2}\right) = \frac{E}{1-\nu}\left(\frac{\partial u}{\partial x} + \frac{\partial v}{\partial y}\right) \quad (3.37)$$

In Eq. (3.37), E is the Young's modulus of the material, $\nu$ is the Poisson's ratio. From Eq. (3.37) some interesting graphical representations of the field can be obtained.

The direction of the $\mathbf{d}(\mathbf{x})$ vector, Eq. (3.3), is given for small rotations and deformations,

$$\theta_d(\mathbf{x}) = \text{arctg} \frac{v(\mathbf{x})}{u(\mathbf{x})} \quad (3.38)$$

In the field of displacements one can define lines of equal inclination of the vector displacement. In [3], these lines are called isoclinic lines of the displacement field.

From [1], the relative displacement vector field is a vectorial field governed by the differential equation:

$$\frac{dy}{dx} = \frac{v(\mathbf{x})}{u(\mathbf{x})} \quad (3.39)$$

The above differential equation defines field trajectories that in [2] are called "trochias". The displacement vectors are tangent to the trochias. Eq. (3.34) gives the relationship between the displacement isoclinics $\Theta_d$ and the isoclinics corresponding to the direction of the principal strains $\theta_{p1}, \Theta_{s1}$.

## Connecting Kinematical Variables and Dynamic Variables

In the preceding sections, a mathematical consistent model for the analysis of kinematical variables based on the fundamental principles of Continuum Mechanics was introduced. This is an important step towards the process of matching mathematical models with experimental observations. The next step is to connect kinematical variables to dynamical variables, completing a process needed both from the theoretical point of view as well as in terms of actual engineering applications. Let us begin with an intuitive approach to this relationship. The Eulerian description of continuum displacements in a uniform field is illustrated in Fig. 3.7.

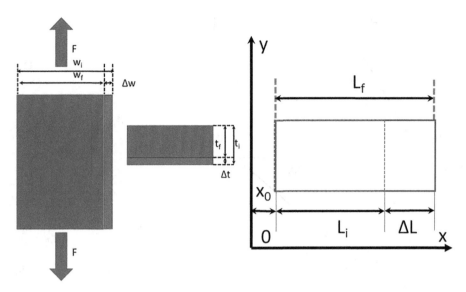

**Fig. 3.7** Tensile specimen subjected to uniform tensile stress, plane stress condition

The longitudinal displacement of a generic point can be expressed by the following equation,

$$u(x) = (x - x_0) \frac{L_f - L_i}{L_f} \tag{3.40}$$

Differentiating Eq. (3.40), it follows,

$$\frac{\partial u}{\partial x} = \frac{L_f - L_i}{L_f} = \varepsilon_x^E \tag{3.41}$$

This is the value given by the Euler-Almansi strain tensor for a constant field, Eq. (3.23).

The strain and the stress tensors depend on the configuration of the medium. In this case, we are working with the deformed configuration, the strain is the Eulerian strain, the stress is the true stress, the force is acting in the deformed configuration, the corresponding stress tensor is the Cauchy stress tensor. In Fig. 3.7, it is shown that since the specimen is pulled in the axial direction, the cross section is changed. The Eulerian stress $\sigma^E$ is given by the applied force divided by the actual cross-section area,

$$A_f = w_f \times t_f \tag{3.42}$$

The stress is,

$$\sigma^E = \frac{F}{A_f} \tag{3.43}$$

The above basic derivation for a tensile specimen field in the context of plane stress field can be generalized for the general 2-D tensor.

As shown in Fig. 3.8a, the parallelogram $M_oN_oM'_oP_o$ in the undeformed configuration after deformation becomes the square $M_1N_1M'_1P_1$. To this square correspond the components of the Eulerian-Almansi strain tensor in the deformed position, Eq. (3.23). As mentioned before, the stress is the true stress and this stress is defined by the Cauchy stress tensor. In Fig. 3.8b, there are shown the three components in 2D of the true stress tensor that correspond to the deformed state. This relationship can be generalized for the six components of the Cauchy stress tensor in 3D. From all the arguments presented in this section, the Euler-Almansi strain tensor is compatible with the Cauchy stress tensor.

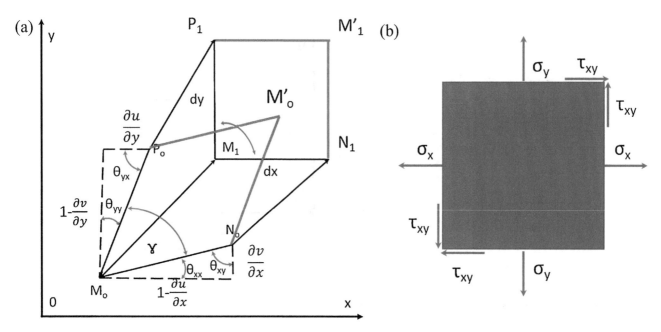

**Fig. 3.8** (a) An element of the surface represented by the parallelogram $M_o N_o M'_o P_o$ in the undeformed position becomes a square in the deformed position. (b) Stress components applied to the element of area

## Constitutive Functions

While the kinematics of the continuum is straightforward, the dynamics is far more complex than the kinematics. The reason for this feature of the dynamics of the continuum is that the dynamics involves the realm of modeling the behavior of materials and covers an extremely wide spectrum of behaviors. To relate the strain and stress tensors, it is necessary to define constitutive functions. Since in our derivations we are considering RVE of different materials, let us introduce some simplifications. Let us assume constitutive functions that are not path dependent, and to further simplify arguments we will assume that the RVE is not affected by the time. Also, let us recall that to be physically valid a constitutive function must be independent of the utilized frame of reference. Among all materials of technical interest, a group of materials that comply with the above conditions are the Cauchy-elastic materials. No real materials comply with the requirements of Cauchy-elastic materials, which is just a mathematical model. However, many materials utilized in engineering applications and biomaterials can be described as Cauchy-elastic materials under certain circumstances. The most general approach to derive a constitutive function is to define for a given medium an internal energy function. This function will contain thermodynamic variables such as temperature, entropy etc. as well as kinematic variables that depend on the selected strain tensor. It should be remembered that constitutive equations depend on an adopted definition of deformation as well as on constants or functions that must be obtained by experiments where the adopted tensor variables can be measured, and applied forces are also measured. The strain energy function can be expressed as a function of the adopted deformation tensor or a strain tensor that has been chosen and at the same time as a function of a measure of stress that is compatible with the adopted strains. If the internal energy $W(E_{ij})$ is defined as a function of a given strain tensor $E_{ij}$, and assuming that the internal energy depends on the deformation gradient only and if one ignores thermodynamic variables and energy dissipation processes, it can be written,

$$\Sigma_{ij} = \frac{\partial W(E_{ij})}{\partial E_{ij}} \tag{3.44}$$

In Eq. (3.44), the tensor $\Sigma_{ij}$ has to be compatible with $E_{ij}$ since we are dealing with a selected configuration of the continuum. An equivalent statement is associated with the adopted strain tensor $E_{ij}$, there is a stress tensor $\Sigma_{ij}$ which is the conjugate of the strain in the definition of virtual work. If we are using the Eulerian description, for example, $\Sigma_{ij}$ is the Cauchy stress tensor and $E_{ij}$ the Euler-Almansi strain tensor. A simple extension of classical elastic mediums is the Saint Venant-Kirchhoff elastic medium defined by a two parameters energy function,

$$W\left(E_{ij}^{E}\right) = \frac{C_{1E}}{2}\left(\text{tr}\left[E_{ij}^{E}\right]\right)^{2} + C_{2E}\left(\text{tr}\left[E_{ij}^{E}\right]\right)^{2} \tag{3.45}$$

In the above equation, $C_{1E}$ and $C_{2E}$ are constants that play the role of the Lamé elastic constants in linear elasticity. From Eq. (3.45), it follows,

$$\Sigma_{ij} = \frac{\partial W\left(E_{ij}^{E}\right)}{\partial E_{ij}^{E}} = C_{1E}\text{tr}\left(E_{ij}^{E}\right) + 2C_{2E}E_{ij}^{E} \tag{3.46}$$

In Eq. (3.46), according to the preceding developments, the strain tensor to be introduced is the Euler-Almansi strain tensor of Eq. (3.24) and the stress tensor is the Cauchy stress tensor,

$$\Sigma_{ij} = \begin{bmatrix} \sigma_x & \tau_{xy} \\ \tau_{xy} & \sigma_y \end{bmatrix} \tag{3.47}$$

Expanding Eq. (3.47), we obtain

$$\sigma_x = C_{1E}\left(\varepsilon_x^{E} + \varepsilon_y^{E}\right) + 2C_{2E}\varepsilon_x^{E} \tag{3.48}$$

$$\sigma_y = C_{1E}\left(\varepsilon_x^{E} + \varepsilon_y^{E}\right) + 2C_{2E}\varepsilon_y^{E} \tag{3.49}$$

$$\tau_{xy} = 2C_{2E}\varepsilon_{xy}^{E} \tag{3.50}$$

## Applications of the Derived Relationships of Continuum Kinematics to Experimental Data of Derivatives

Having presented basic conclusions arrived at in Continuum Mechanics mathematical models for the deformations of solids, let us now verify these derivations with actual experimental data.

The tensor of the derivatives computed in a region of a large deformations and large rotations given in [1] is,

$$[J] = \begin{bmatrix} \frac{\partial u}{\partial x} = 0.06112 & \frac{\partial u}{\partial y} = 0.06112 \\ \frac{\partial v}{\partial x} = 0.1441 & \frac{\partial v}{\partial y} = 0.1441 \end{bmatrix} \tag{3.51}$$

Then, according to Eq. (3.51):

$$\frac{\partial u}{\partial x} + \frac{\partial v}{\partial y} = 0.06112 + 0.1441 = 0.20522 \tag{3.52}$$

Furthermore, since in [1], $\varepsilon_{p1}^{E} = 0.2169$ and $\varepsilon_{p2}^{E} = -0.02485$,

$$\varepsilon_{p1}^{E} + \varepsilon_{p2}^{E} = 0.2169 - 0.0248 = 0.1991 \tag{3.53}$$

The difference between the two values is 6%. This difference can be considered within the margins of experimental error.

## Experimental Observations Related to the Concept of RVE

Microscopic patterns corresponding to a metallic particulate composite [1, 5] illustrate the scale transition in connection with the concept of RVE. The four images of Fig. 3.9 reproducing figures of Ref. [1] correspond to a tensile specimen at two different scales, millimeter scale, micron scale. Although patterns at each scale are very different, it is possible to verify a connection between these images. The properties of the tensile specimen are well documented; finite element (FE) study [6] of the composite was performed and mechanical properties were computed.

Figure 3.10 shows two strain-stress curves of the composite, one corresponds to the tensile machine recording, the other to the FE results [6]. The results of the analysis of the tensile specimen analysis are summarized in Table 3.1. Figure 3.10 shows that up to certain level of stress the composite behaves as a quasi-elastic material and from the point of view of patterns at the millimeter scale shown in Fig. 3.9a, the fringes correspond to a uniformly loaded tensile specimen.

At the micron scale the field of displacements is extremely complex, Fig. 3.9a. To analyze these images the derivatives of the projected displacements were determined. The computed values were smoothed using a filter in order to limit peak values that correspond to the presence of dislocations (areas of material discontinuities).

In Table 3.1, (C1) corresponds to the value of the tensile stress $\sigma$ (see Fig. 3.10 and [6]), (C2) is the corresponding strain. (C3) is the secant modulus, quotient of (C1) and (C2). (C4) Poisson's ratio, is the ratio between (C6) and (C5). (C5) is the average of the Eulerian strain in the y-direction $\varepsilon^E_{yavg}$ obtained from the V pattern of Fig. 3.9a. Column (6) is the average of

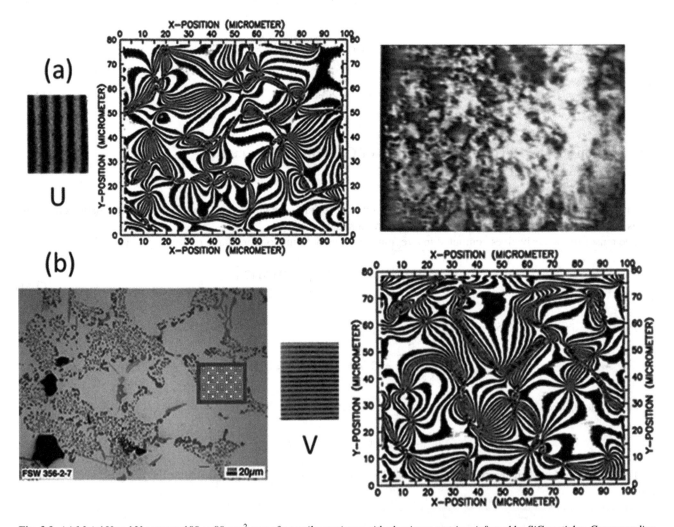

**Fig. 3.9** (a) Moiré U and V patterns $100 \times 80\ \mu m^2$ area of a tensile specimen with aluminum matrix reinforced by SiC particles. Corresponding U and V patterns in the mm scale and microphotography of the observed region, black areas SiC particles, bright areas aluminum matrix; (b) Metallographic image of the aluminum matrix and the observed region of $100 \times 80\ \mu m^2$. The observed region is smaller than the average grain size

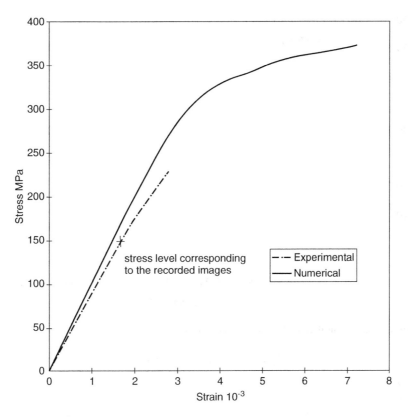

**Fig. 3.10** Stress-strain curve of the Al-SiC tensile specimen

**Table 3.1** Summary of main information concerning the Al-SiC composite test

| σ (C1) (Fig. 3.10) | $\varepsilon_y$ (C2) (Fig. 3.10) | $E_c$ (C3) Secant mod. | $\nu_c$ (C4) Poisson's ratio | $\varepsilon^E_{yavg}$ (C5) Field avg. | $\varepsilon^E_{xavg}$ (C6) Field avg. | Error on $\varepsilon_y$ (C7) |
|---|---|---|---|---|---|---|
| 149.38 MPa | $1.675 \times 10^{-3}$ | 89.2 MPa | 0.39031 | $1.651 \times 10^{-3}$ | $0.645 \times 10^{-3}$ | 1.4% |

the Eulerian strain in the x-direction $\varepsilon^E_{xavg}$ obtained from the U pattern of Fig. 3.9a. (C7) is the percent difference between (C2) and (C5). The two values of $\varepsilon_y$, the one coming from Fig. 3.10 and that of the average $\varepsilon^E_{yavg}$ match very closely, only 1.4% difference. This result indicates that the field average at a lower scale $\varepsilon^E_{yavg}$ provides the value of the strain $\varepsilon_y$ at the upper scale.

## Experimental Verification of Constitutive Equations Relationships

An example taken from [4] (Figs. 3.13 and 3.14 are also reproduced in [3]) will be utilized to ascertain the validity of the constitutive equations of the RVE derived in section "Constitutive Functions". A disk under diametrical compression was utilized and to this disk were applied moderately large deformations. A plate of a urethane rubber was cast and from this plate the tested disk and two tensile specimens were machined. The diameter of the disk is four inches (10.16 cm), the thickness is t = 0.5 in. (t = 1.27 cm). The initial cross section $A_o = 4 \times 0.5 = 2$ sq. in. A 300 lines/inch grating (pitch = 84.67 microns) was photo-engraved in the disk. The tensile specimens were standard tensile specimens with a cross-section of 0.629 sq. in. A 300 lines/inch gratings was photo-engraved in the tensile specimens, in the longitudinal direction in one of them and in the transversal direction in the other.

The corresponding graph of the results obtained from the tensile specimen utilized to get the constant $C_{1E}$ are shown in Fig. 3.11. It can be seen that $C_{1E} = 479$ psi ($C_{1E} = 3.303$ MPa).

The results of the second tensile specimen are shown in Fig. 3.12. It can be seen that $C_{2E} = 993.39$ psi ($C_{2E} = 6.852$ MPa). The ratio of the two constants is $r_c = 0.482$, if $r_c = 0.5$ the material is incompressible.

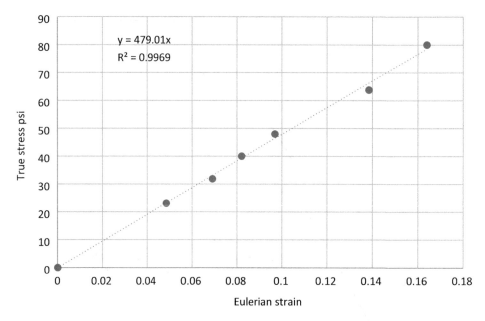

**Fig. 3.11** Results of the tensile specimen utilized to determine the constant $C_{1E}$

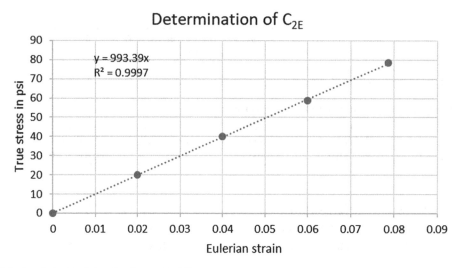

**Fig. 3.12** Results of the tensile test to determine the constant $C_{2E}$

Figure 3.13 shows the moiré patterns of the compressed disk. The moiré fringes were produced by superposition of the undeformed and the deformed images in a photographic plate.

For the purpose of comparing the different results represented in Fig. 3.13, it is necessary to utilize a dimensionless representation. The unloaded disk has a circular shape, upon loading becomes quasi-elliptical with the principal axis remaining axis of symmetry. In the plot of Fig. 3.14, the horizontal axis reports the x coordinate divided by the radius of the deformed disk, the dimensionless radius is then equal to $r = 1$.

The stresses are also dimensionless in the plot of Fig. 3.14. Stress is normalized by dividing the actual stress obtained from the constitutive equations by the average stress $\sigma_{avg}$. The average stress is determined by the numerical integration of the stress plots. The theoretical curve corresponds to the classical elasticity solution of the disk under diametrical compression. The Lagrangian strain was obtained by applying the equation,

$$\varepsilon_x^L = \frac{\varepsilon_x^E}{1 - \varepsilon_x^E} \tag{3.54}$$

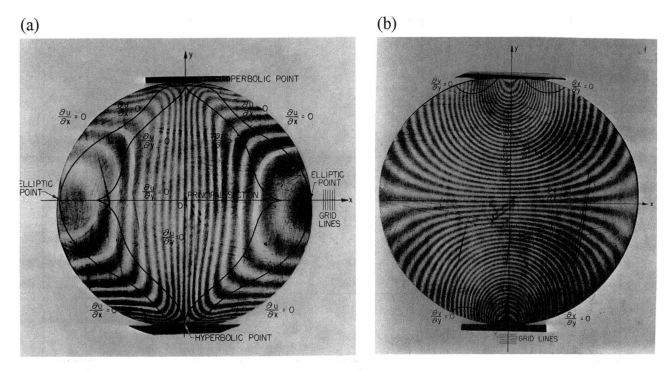

**Fig. 3.13** Disk under diametrical compression, (**a**) u-pattern, (**b**) v-pattern

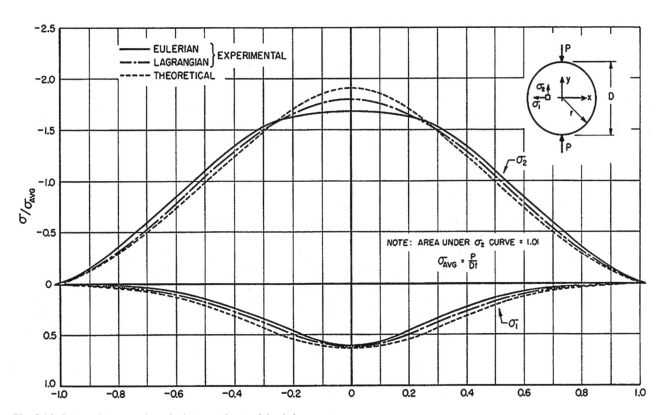

**Fig. 3.14** Principal stresses along the horizontal axis of the disk

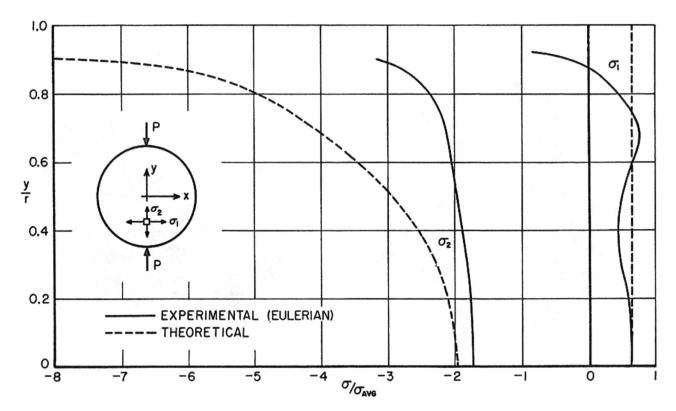

**Fig. 3.15** Principal stresses along the vertical axis of symmetry of the disk

The average of the dimensionless Eulerian curve was 1.01 indicating an error of 1% in the condition of static vertical equilibrium.

The principal stresses along the vertical symmetry axis are plotted in Fig. 3.15. It can be seen that the principal stress $\sigma_2$ of the theoretical solution differs considerably from the $\sigma_2$ since the theoretical solution utilizes a singularity solution assuming a point loading while the experimental solution corresponds to a load resulting from the contact stresses between the rubber and the steel piece utilized to apply the load. The stress $\sigma_1$ is for the most part of the depth positive but near the contact becomes negative.

To verify that the Euler-Almansi strain tensor and the Cauchy the stresses tensor corresponds to each other it is necessary to verify that the obtained stresses corresponding to the RVE from the constitutive Eq. (3.48) through (3.50) satisfy the static equilibrium conditions. In the horizontal direction, the condition of zero resultant force is satisfied by the symmetry of the patterns of the principal stresses $\sigma_1$. In the vertical direction at different depths, the integrated curves of $\sigma_2$ satisfy equilibrium with maximum differences less than 2%. By analyzing the region near the contact of the urethane rubber disk and the steel bar utilizing a larger scale, it is possible to obtain the stress distribution in this area. At the depth of 0.10r, Fig. 3.16, there is a point where $\sigma_x = 0$ and the $\sigma_x$ becomes compressive in the contact region of approximately width w = 0.075r. In a smaller region of the order of 0.05r, the vertical stresses $\sigma_y$ are distributed and the lever arm is approximately 0.025r.

With this information we can verify if the sum of the moments defined by the equilibrium of a quarter of the disk is equal to zero. In order to compute the moments that define the equilibrium of the quarter of the disk, positive moments are assumed clock-wise. The computation of the vertical moment $M_v$, involves the resultant force R, the force P denoting the force applied to the disk, and L that represents half of the contact stresses is L = −0.500 P. Also, one must remember the dimensionless representation of the vertical and horizontal coordinates that make the radiuses r = 1. In the case of the vertical axis, the dimensionless coordinates y is obtained by dividing the corresponding y-values by the length of the minor axis of the deformed disk. In the following equation, the symbol r is utilized to indicate that we have the dimensions of a force times a length, a moment. The moment of the vertical forces $M_v$ is given by,

$$M_v = R_v \times 0.3r - L \times 0.025r = 0.139 \, Pr \tag{3.55}$$

**Fig. 3.16** Stress distribution of the quarter of the disk to determine the moment equilibrium conditions

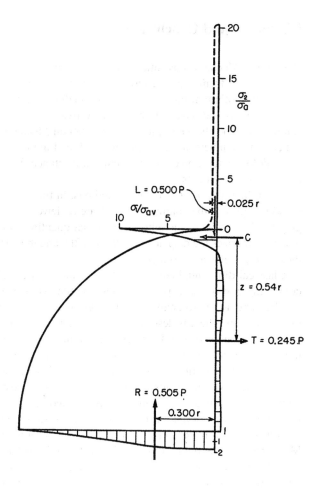

The moment $M_h$ of the horizontal forces, taking into consideration that in view of the equilibrium of horizontal forces $C = -T = 0.245\,P$,

$$M_h = -0.245P \times 0.54\ r = -0.1323\,Pr \tag{3.56}$$

The sum of the moments is,

$$\Sigma_M = 0.0067\,Pr \tag{3.57}$$

The sum of the moments becomes zero to approximately two significant figures. Then the sum of the moments, that should be zero, has referred to the $M_v$ an error equal to 1.7%.

From the preceding developments, one can conclude that, within the experimental error, the Euler-Almansi strain tensor and the Cauchy stress tensor are conjugate with each other in the definition of virtual work. This a very important result in the application of techniques that provide displacements and displacements derivatives. It is interesting to remind the fact that has been pointed out in [3]. When utilizing the Lagrangian description, the displacements and their derivatives should be measured along lines that initially were parallel to the Lagrangian coordinated axes, but in the deformed condition are no longer aligned with the coordinate axes. While in the Eulerian description, the displacements and their derivatives must be determined along lines that are parallel to the Eulerian coordinate axis. This fact is illustrated in Fig. 3.8, which associates the Euler-Almansi strain tensor with the Cauchy stress tensor. In this section, the above pointed relationship has been applied to a given type of constitutive equations and proved that satisfies essential conditions, the static equilibrium conditions.

## Discussion and Conclusions

Continuum Mechanics assumes that mathematical models explaining interactions between materials and applied external inputs can be formulated in terms of a fundamental hypothesis: materials in their interactions with their surroundings behave as continuum media. Continuum media in turn can be mathematically represented by mathematical functions that are continuous and have continuum derivatives at least up to the third order derivatives. The validity of the mathematical solutions can be checked utilizing Experimental Mechanics. This process began with the development of Photoelasticity at the end of the nineteenth century and continued in the twentieth century. In the twentieth century, an explosion of different methodologies took place. Of particular importance from the point of view of full field observations are the methods that measure displacements.

From the point of view of these methods, in this paper, formulas connected to large deformations and large rotations of the kinematics of the continuum in 2D are reviewed. These basic formulas are connected with the processes of converting gray level recordings in a light sensor, scalar quantities into vectorial fields. The different components that define the phasor that characterizes the displacement field in 2D, and the complex field mathematics that makes possible the determinations of these components are discussed.

Classical differential equations of the solution of 2D kinematics of the continuum are related to the general process of converting gray levels as potentials defining the 2D displacement fields.

The next step is to analyze the tensors of the derivatives of the displacement functions in the Eulerian description of the continuum. Then, is described the connection between these derivatives and the differential geometry plot that provides the local behavior of the deformed continuum in the Eulerian description. This differential geometry description leads to the Euler-Almansi strain tensor. Useful and interesting properties of this tensor are derived. The next topic dealt with is the concept of constitutive equations connected with the Eulerian description of the continuum. The application of the Hill-Mandel condition leads to the concept of constitutive equation through the definition of energy contained elementary volume. A particular type of material is selected, the Saint-Venant-Kirchhoff elastic medium to illustrate the connection of the Euler-Almansi strain tensor and the Cauchy stress tensor. The theoretical mathematical models presented in the paper are illustrated with experimental verifications that give a link between the points of view of theoreticians of Continuum Mechanics and the point of view of Experimental Mechanics practitioners.

At the basis of the developments presented in this paper is the concept of RVE, a bridge between the actual discontinuous nature of material media and their corresponding discontinuous structure. There are different disciplines connected with the mechanics of solids, Theoretical Mechanics of the Continuum, with two branches solution via selected functions, or numerical solutions, for example finite elements. Another discipline is Experimental Mechanics that allows the verification of theoretical or numerical results. A third discipline is the study of actual atomic configurations and the force interactions between atoms. These disciplines have evolved in time to give an answer to a centuries old question: why things break and fail? The corresponding counterpart of this question is how to control the corresponding processes by manufacturing materials and given them corresponding safe shapes. An alternative to the RVE is the statistical volume element (SVE), that in FE is referred to as stochastic volume element. While RVE utilizes statistical averages, SVE deals directly with solutions obtained applying the Theory of Random Functions to the continuum thus arriving to the theory of continuum random fields.

In the process of interaction of these disciplines to arrive to the pointed out safe solution Experimental Mechanics plays a fundamental role both as a tool to verify theoretical predictions and to generate new ideas and concepts to create models that can provide practical answers to the above posed questions. The content of this paper may give some partial answers to some of these questions utilizing the RVE concept. This paper brings some insights on the role of Experimental Mechanics as source of future developments in the search of answers to the old question: Why things break and fail?

## References

1. Sciammarella, C. A., Lamberti, L., & Sciammarella, F. M. (2019). The optical signal analysis (OSA) method to process fringe patterns containing displacement information. *Optics and Lasers in Engineering, 115*, 225–237.
2. Sciammarella, C. A., & Lamberti, L. (2015). Basic models supporting experimental mechanics of deformations, geometrical representations, connections among different techniques. *Meccanica, 50*, 367–387.
3. Durelli, A. J., & Parks, V. J. (1970). *Moiré analysis of strain*. Englewood Cliffs, NJ: Prentice-Hall, Inc.
4. Sciammarella C.A. (1960). *Theoretical and experimental study on moiré fringes*. PhD dissertation, Illinois Institute of Technology, Chicago (USA).
5. Sciammarella, C. A., Sciammarella, F. M., & Kim, T. (2003). Strain measurements in the nanometer range in a particulate composite using computer-aided moiré. *Experimental Mechanics, 43*, 341–347.
6. Sciammarella, C. A., & Nair, S. (1998). Micromechanics study of particulate composites. In *Proceedings of the SEM Spring Conference on Experimental Mechanics* (pp. 188–189).

# Chapter 4
# Study the Deformation of Solid Cylindrical Specimens Under Torsion Using 360° DIC

Helena Jin, Wei-Yang Lu, Jay Foulk, and Jakob Ostien

**Abstract** In this work, the plastic deformation and ductile failure of solid Al 6061-T6 and 304L tube specimens under torsion loading was investigated using a new developed 360° DIC system. The tube specimens of both ductile metals can exhibit very large rotation under torsion loading, which makes it very difficult to experimentally measure the deformation at failure. In our earlier work, 3D DIC system composed of one pair of cameras was applied to investigate the local deformation. The DIC system with one pair of cameras was only able to track the area of interest within a rotation angle of 90° or smaller. However, the specimens failed at rotation angles larger than 240°. The DIC system with one pair of cameras was difficult to obtain the critical strain data at large rotation. In this work, a new 360° DIC system which consists of four pairs of cameras was designed and incorporated into the experiment so that the whole surface can be imaged during the rotation. The new DIC system was able to track the area of interest and measure the deformation till the specimen failure. Shear strain at failure location can be measured and related to the global loading condition.

**Keywords** Digital image correlation · Shear loading · Large deformation

## Introduction

Fracture and failure in ductile metals is a complex issue, especially under shear dominated loading. There are numbers of efforts in numerical analysis and modeling of ductile fracture and failure, but very few experimental efforts. We have previously studied the ductile failure of Al 6061 under various combination of tension and torsion using thin wall tubular specimens. In this work, we conducted torsion experiments of solid cylinder specimens to introduce very large shear deformation. The traditional 3D-DIC technique with one pair of cameras was initially applied to measure the localized deformation at the shear band, but it faces difficulty of tracking the same area of interest due to large rotation angle. A new 360° DIC system which consists of four pairs of cameras was designed for the torsion experiments to measure the deformation around the whole cylinder specimen and track the initiation of failure. This paper will discuss the multiple challenges facing the new DIC system.

## Specimen and Experiment

Two types of materials are of interest in this work—Al 6061 and stainless steel 304L. The experiment was conducted using MTS table top Bionix system with both torsion and tension/compression capabilities. The specimens were gripped using the hydraulic grip at both ends. During the experiment, the axial force and displacement, as well as the rotation angle and torque were monitored. The specimen was maintained under zero axial force. The specimen was painted with black and white paint to generate speckle patterns for the DIC technique to measure the full-field deformation. Figure 4.1 shows the experimental setup with cylinder specimen in the middle and four pairs of cameras facing the specimen. Each pair of cameras have their own LED light facing the specimen. Each pair of cameras were able to capture 90° of the specimen surface and four pairs of cameras captured the whole circumference of the specimen. Each pair of cameras need to be calibrated separately first to have accurate measurement in their own field of view. They are firmly mounted to the tripod to avoid relative motion between two

**Fig. 4.1** Torsion experiment setup with cylinder specimen in the middle and four pairs of cameras facing the specimen. 1 specimen, 2–5 DIC camera pairs and LED lights

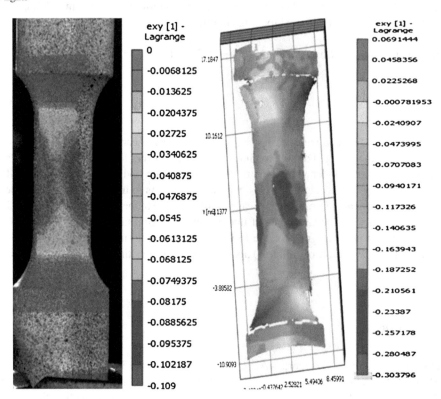

**Fig. 4.2** (**a**) DIC results from one system; (**b**) combined DIC results from multi-systems after transformation

cameras once they were calibrated. Meanwhile, all four pairs of cameras need to be secured tightly to avoid relative motion among each pair once they were set up and calibrated.

## Analysis

To obtain the full-field displacement and strain from the DIC technique, two challenges need to be overcome: (1) Multi-View registration for the four pairs of DIC systems. (2) Tracking the same area in different DIC systems as the specimen underwent large rotation. As shown in Fig. 4.2a, the full-field deformation was first calculated from each individual DIC system in its own coordinate system. Then one system was selected as the reference system. The other systems were transformed to the reference system using the transformation tensor that was calculated from a common set of rigid body motion. The cylindrical view of the specimen can then be obtained based on the same coordinate system. Figure 4.2b shows the cylindrical whole field view after coordinate transformation.

## Conclusions

The new 360° DIC system was a powerful tool to obtain the full-field deformation around the whole cylindrical view of the specimen. It enabled us to understand the initiation of the ductile fracture and failure when the specimens underwent large shear deformation. However, challenges remained ahead as we aimed to obtain the deformation when the specimens rotated through views of multiple DIC systems.

**Acknowledgement** Sandia National Laboratories is a multimission laboratory managed and operated by National Technology and Engineering Solutions of Sandia, LLC., a wholly owned subsidiary of Honeywell International, Inc., for the U.S. Department of Energy's National Nuclear Security Administration under contract DE-NA-0003525. This paper describes objective technical results and analysis. Any subjective views or opinions that might be expressed in the paper do not necessarily represent the views of the U.S. Department of Energy or the United States Government.

# Chapter 5
# Multiscale XCT Scans to Study Damage Mechanism in Syntactic Foam

Helena Jin, Brendan Croom, Bernice Mills, Xiaodong Li, Jay Carroll, Kevin Long, and Judith Brown

**Abstract** In this work, we applied the in-situ X-ray Computed Tomography (XCT) mechanical testing method that coupled the in-situ mechanical loading with the XCT imaging to study the damage mechanism of GMBs inside the Sylgard as the material was subject to mechanical loading. We studied Sylgard specimens with different volume fraction of GMBs to understand how they behave differently under compression loading and how the volume fraction of GMBs affect the Sylgard failure.

Both high resolution (1.5 μm/voxel) and low resolution (10 μm/voxel) XCT imaging were performed at different loading levels to visualize the GMB collapse during the compression of Sylgard with different volume fraction of GMBs. Feret shape of GMBs were calculated from the high resolution XCT images to determine whether the GMBs were intact or fractured, as well as the relationship between the size distribution of GMBs and their Feret shapes. Through these quantitative analysis of the high resolution XCT data, we were able to understand how the size and volume fraction of GMBs affect their failure behavior. The Digital volume correlation (DVC) technique was applied to the low resolution XCT images to calculate the local deformation of Sylgard specimen, which enabled us to understand the different failure propagation and failure mechanisms of Sylgard with different volume fraction of GMBs.

**Keywords** Sylgard · Glass microballoons (GMB) · X-ray computed tomography · Digital volume correlation (DVC)

## Introduction

Syntactic foams are lightweight polymeric composites reinforced with hollow glass microballoon (GMB) particles. A key outstanding challenge in the design and modeling of syntactic foams is predicting the onset and duration of GMB collapse, as these two parameters govern the energy absorption of the foam. At the heart of the issue is confusion over the precise mechanisms that lead to damage initiation and propagation due to complex dependency on GMB geometry and mechanical properties. Changes in the syntactic foam microstructure due to GMB diameter, wall thickness, volume fraction and interfacial characteristics can have pronounced effects on the mechanical response and onset of damage. Moreover, the performance of syntactic foams is intrinsically stochastic, since damage initiation depends on the random packing of individual GMBs. This has motivated the use of in situ X-ray Computed Tomography (XCT) mechanical testing to visualize the 3D foam microstructure and damage behavior. In situ XCT allows for quantitative measurement of individual GMBs and their spatial arrangement, coupling with Digital Volume Correlation (DVC), enabled us to simultaneously visualize the syntactic foam microstructure and quantify the local deformation and damage in the composite and find the relationship between these two.

---

H. Jin (✉) · B. Mills
Sandia National Laboratories, Livermore, CA, USA
e-mail: hjin@sandia.gov; mills@sandia.gov

B. Croom · X. Li
University of Virginia, Charlottesville, VA, USA
e-mail: bpc2nr@virginia.edu; xl3p@virginia.edu

J. Carroll · K. Long · J. Brown
Sandia National Laboratories, Albuquerque, NM, USA

## Experiment and Results

*In situ* compression experiments (Fig. 5.1a) of the syntactic foam specimens were performed inside X-Radia 520 Versa machine. Tomograms were acquired at the original undeformed state as well as at global engineering strain increments of 7% up to 50% compression at two resolutions (high resolution of 1.7 μm per voxel, and low resolution of 10 μm per voxel) to simultaneously image the collapse of GMBs with high detail, and also capture the macroscopic global deformation. Special care was taken to align the high-resolution tomograms at different loading steps so that the behavior of individual GMBs from the same set could be tracked.

The high-resolution tomograms with 1.7 μm per voxel resolution captured the detailed morphology of the damaged GMBs (Fig. 5.2a), allowing segmentation of the glass walls from the rubber matrix and identification of the specific damage mechanisms. Two types of GMB failure were identified. The most common failure mode was fracture of the GMB walls, which created several glass fragments that were visible in the images. Second, a trace fraction of GMBs were observed to delaminate from the matrix. Figure 5.2b shows the 3D renderings of the GMBs. Standard image processing techniques were implemented in 3D using Avizo 9.0 software to denoise, segment, label and measure individual GMBs. The GMBs were classified into *intact* and *fractured* based on Feret shape, which provided a 3D measure of the GMB aspect ratio. Thresholding of the Feret shape (where FS > 1.5) identified collapsed GMBs. Analysis of these trends showed continued GMB collapse over

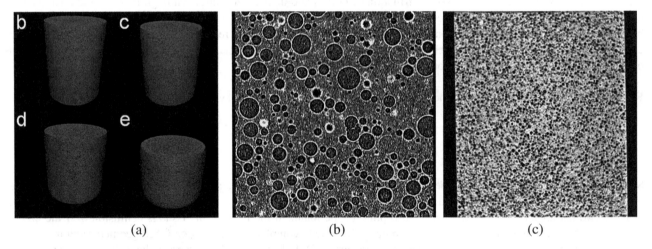

**Fig. 5.1** (**a**) In-situ XCT compression experiment of Sylgard, (**b**) high-resolution of Sylgard with GMBs, (**c**) low-resolution of Sylgard with GMBs

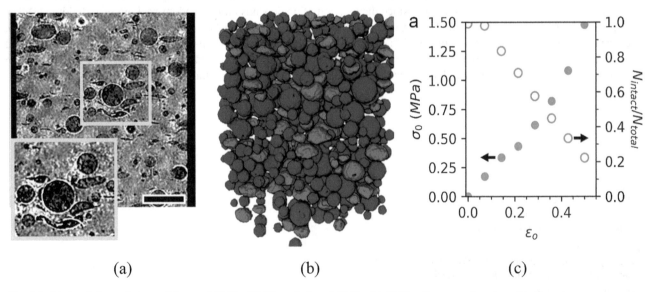

**Fig. 5.2** (**a**) detailed morphology of damaged GMBs, (**b**) 3D rendering of GMBs, (**c**) GMB collapse as a function of load

a range of compression from 14% to beyond 50%. Quantitative image analysis of the high-resolution tomograms provided measurements of GMB collapse as a function of load (Fig. 5.2c).

**Acknowledgements** Sandia National Laboratories is a multimission laboratory managed and operated by National Technology and Engineering Solutions of Sandia, LLC., a wholly owned subsidiary of Honeywell International, Inc., for the U.S. Department of Energy's National Nuclear Security Administration under contract DE-NA-0003525. This paper describes objective technical results and analysis. Any subjective views or opinions that might be expressed in the paper do not necessarily represent the views of the U.S. Department of Energy or the United States Government.

# Chapter 6
# An Investigation of Digital Image Correlation for Earth Materials

Nutan Shukla and Manoj Kumar Mishra

**Abstract** Strain measurement is vital in material and mechanical testing. The conventional contact based approaches as strain gauges, extensometers, acoustic emission, seismic waves etc. have limitations like sensitivity to contact areas, susceptible to external noise, vulnerable to breakage, influenced by surrounding environment etc. Recently, development of noncontact based methods has eliminated these challenges. Digital Image Correlation is an optical, noncontact based technique which is being widely used in measuring full-field displacement. The technique is based on acquisition of digital images taken before and during loading and developing a correlation between them with subpixel accuracy. In this paper, the behavior of equivalent coal specimens at varying depth from 92 to 126 m have been evaluated for static loading condition using non-contact based approach i.e. Reliability-guided Digital Image Correlation technique (RG-DIC). There exists a good agreement between the contact (strain gauge) and noncontact based approaches (DIC) for static loading condition. So, Digital Image Correlation (DIC) can also be a reliable strain measuring approach for investigating for earth materials.

**Keywords** DIC · RG-DIC · Strain measurement · Static loading · Full-field displacement

## Introduction

Strain measurement is an important tool to evaluate the behavior of any material for its both short and long term usage. Experimental strain analysis has gained popularity in manufacturing industries because of its ability to test the durability of the specimen. However, measuring the parameters outside the laboratory, in real cases, requires a tough choice among conventional methods keeping accuracy, simplicity and cost balanced. The conventional methods are mostly contact based methods that include electrical strain gauges, mechanical dial gauges, acoustic emission techniques etc. The contact based methods have issues like sensitivity to noise, damage to the recording unit, loss of contact of sensors with the surface, influence of environmental noise and temperature, measurement error etc. [1, 2]. Non-contact based methods are categorized as interferometric technique and non-interferometric technique which precede contact based methods in terms of accuracy, human error and repeatability of experiment. DIC technique is evolving to be a very effective strain measuring tool for continuous material [3].

In this work, a non-contact based optical method Digital Image Correlation (DIC) [4–6] has been used that acquires an image of the object, stores it and performs image correlation analysis to extract full field deformation of that object. The concept involves comparing the gray intensity changes of the object surface in the acquired images with the reference image. This technique has become widely popular in measuring full field deformation in various industries viz. automobile, solar wind power, medical etc. [7–10].

Optimizing the underground coal pillars dimension is very important to maximize the extraction without compromising the safety of the system. This involve investigation into the deformation behavior of the coal pillar at varying load. So far there is limited data available on the efficiency of DIC technique on earth materials.

---

N. Shukla (✉) · M. K. Mishra
Department of Mining Engineering, National Institute of Technology, Rourkela, India
e-mail: 515mn1001@nitrkl.ac.in

## Sample Preparation

Model pillars have been fabricated with sand-cement to represent coal pillars with comparable strength values. It has been done to eliminate influence of multiple weaknesses in the deformation behavior investigation. The coal specimens were extracted from a depth of 92 to 126 m from an operating underground mine located between latitude 20°55′00″ and 21°00′00″ and longitude 85°05′00″ and 85°10′00″ from MCL, Talcher, India. The uniaxial compressive strength (UCS) test was carried out and the values obtained were between 15 and 33 MPa. The equivalent samples were prepared after many trial and error experiments with varying sand, cement and water ratios. The UCS values of the equivalent specimens obtained are as below.

The prepared model specimens have the UCS values that compares favorably with that of coal samples. Hence, the modeled specimens have been used in further experiments. The model specimens are made of ordinary portland cement and river sand (−0.5 mm mesh). Proper care was taken to achieve uniformity of the cube. The exact amount of sand and cement as per the defined ratio were dry mixed thoroughly and a fixed amount of water (cement: water = 0.42) was mixed to achieve consistency. This mixture was poured in the mould of 10 cm cube and shaked at 3000 rpm to eliminate any void spaces. The moulds were removed after 24 h and specimens were cured at 99% humidity for effecting the chemical reactions for 7 days (Fig. 6.1). The cube specimens were brushed to remove any loose surface particles.

The generation of speckle pattern on the surface is very important in DIC investigations. White and black spray paints were applied on the three sides of the cube to create the speckle pattern (Fig. 6.4). Strain gauge (350.8 ± 0.1 Ω) were fixed on the other side of the cube (Fig. 6.2). The results obtained from those are reported here.

## Experimentation

The prepared models were loaded for UCS values in a servo controlled machine (CCTM, 100T, make: HEICO, India) at a loading rate from 0.5 to 1.0 MPa/s [11]. The entire loading process till fracture was captured by three image acquisition units in three sides in addition to the data generated by the strain gauge from the other side. The entire setup was well illuminated by illuminating lights (Fig. 6.3). The image acquisition unit has a resolution of 1920 × 1080 pixels and 3:2 aspect ratio. The reference and deformed images are given in Fig. 6.4.

The tests that produced a few irregular failure pattern were not considered. The one which exhibited both strain gauge and DIC outputs are discussed here. A total number of 16,368 frames were extracted, out of which 165 frames were taken into consideration for further analysis. An open source 2D digital image correlation MATLAB program Ncorr [12] was used for the analysis which used Reliability Guided DIC algorithm. The strain $\epsilon_{xx}$ and displacement x and y profile are given in Figs. 6.5 and 6.6. A total strain of 2.484% was observed using the Digital Image Correlation technique.

**Table 6.1** Uniaxial compressive strength of modeled specimens

| Ratio | Compressive strength (MPa) |
|---|---|
| 1:1.5 | 27–31 |

**Fig. 6.1** Prepared cube samples (at 28 days)

**Fig. 6.2** Strain gauge installed on a specimen

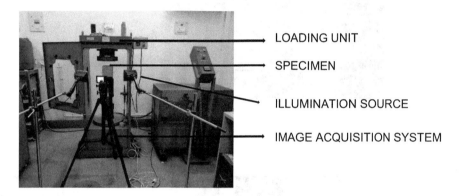

**Fig. 6.3** Laboratory setup

The strain gauges readings are converted to percentage change. In the instant case, the strain gauge reading went up to 2.62%. The stress-strain behavior of the specimen obtained in contact based method and non-contact based method is shown below (Fig. 6.7).

It is observed that both DIC and strain gauge readings were very close to each other in the initial loading phase when specimen offered maximum resistance (±4.6%). There is significant though of lesser magnitude (±9.7%) change seen in the linear range of the loading. However, the difference is wider towards the peak loading where the specimen undergoes irreversible physical disintegration (±21.4%). The contact based method and noncontact based method were analyzed on the equivalent specimen and the strain obtained in both the method are well matched and correlated. An average error of 12% was observed in mapping strain in both the methods.

**Fig. 6.4** (**a**) Reference image, (**b**) deformed image

**Fig. 6.5** Strain $\epsilon_{xx}$ profile

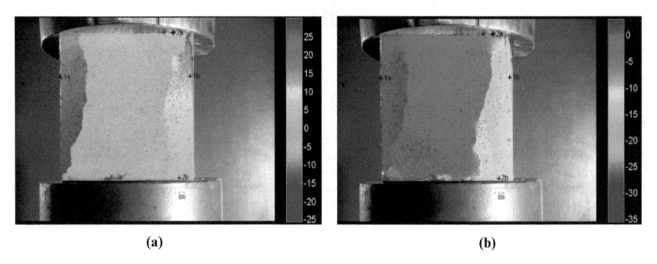

**Fig. 6.6** (**a**) X displacements profile, (**b**) Y displacements profile

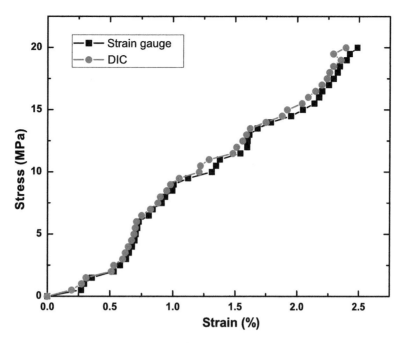

**Fig. 6.7** Stress-strain behavior of a sand cement specimen using DIC and strain gauges

## Conclusion

The investigation was carried out on an equivalent material of coal, an earth material. The Reliability Guided Digital Image Correlation (RG-DIC) algorithm was used to measure the strain under loading condition. A strain of 2.484% was achieved from Digital Image Correlation technique and the same specimen was investigated using strain gauge, a contact based way of measurement. A good correlation between them has been found with a deviation of 12% has been observed between the strain mapping methods.

## References

1. Gale, W. J., Heasley, K., Iannacchione, A., Swanson, P., Hatherly, P., & King, A. (2001). Rock damage characterization from microseismic monitoring. In *Proceedings of the 38th US Rock Mechanics Symposium,* Washington, DC; 7–10 July (pp. 1–313), 2001.
2. Cai, M., Kaiser, P., Tasaka, Y., Maejima, T., Morioka, H., & Minami, M. (2004). Generalized crack initiation and crack damage stress threshold soft brittle rock masses near underground excavations. *International Journal of Rock Mechanics and Mining Sciences, 41*(5), 833–847.
3. Yang, Y., Zhang, F., Hordijk, D. A., Hashimoto, K., & Shiotani, T. (2019). A comparative study of acoustic emission tomography and digital image correlation measurement on a reinforced concrete beam. In *Life-cycle analysis and assessment in civil engineering: Towards an integrated vision* (pp. 2419–2126).
4. Bhattacharjee, S., & Deb, D. (2016). Automatic detection and classification of damage zone(s) for incorporating in digital image correlation technique. *Optics and Lasers in Engineering, 82,* 14–21.
5. Zhou, P., & Goodson, K. E. (2001). Sub pixel displacement and deformation gradient measurement using digital image/speckle correlation. *Optical Engineering, 40*(8), 1613–1620.
6. Acciaioli, A., Lionello, G., & Baleani, M. (2018). Experimentally achievable accuracy using a digital image correlation technique in measuring small-magnitude (<0.1%) homogeneous strain fields. *Materials, 11,* 751.
7. Alan, B., Nik, R., & Claire, D. (2009). Towards a practical structural health monitoring technology for patched cracks in aircraft structure. *Composites Part A: Applied Science and Manufacturing, 40*(9), 1340–1352.
8. Reddy, K. C., & Subramaniam, K. V. L. (2017). Experimental investigation of crack propagation and post-cracking behaviour in macrosynthetic fibre reinforced concrete. *Magazine of Concrete Research, 6*(9), 467–478.
9. Winder, R. J., Morrow, P. J., McRitchie, I. N., Bailie, J. R., & Hart, P. M. (2009). Algorithms for digital image processing in diabetic retinopathy. *Computerized Medical Imaging and Graphics, 33*(8), 608–622.
10. Elmahdy, A., & Verleysen, P. (2018). The Use of 2D and 3D high-speed digital image correlation in full field strain measurements of composite materials subjected to high strain rates. *Proceedings, 2*(8), 538.
11. Standard method of test for elastic moduli of rock core specimens in uniaxial compressions. *American Society for Testing and Materials, ASTM Designation D 3148–72.*
12. Blaber, J., Adair, B., & Antoniou, A. (2015). Ncorr: Open-source 2D digital image correlation Matlab software. *Experimental Mechanics, 55*(6), 1105–1122.

# Chapter 7
# Dynamics of Deformation-to-Fracture Transition Based on Wave Theory

Sanichiro Yoshida, David R. Didie, Tomohiro Sasaki, Shun Ashina, and Shun Takahashi

**Abstract** This paper discusses fracture from the viewpoint of wave dynamics derived from a recent field theory. Based on a fundamental physical principle, the field theory describes deformation and fracture on the same basis. It characterizes deformation as a wave phenomenon where the spatiotemporal oscillatory behavior of the displacement field initiated by an external load is transferred through the material as a sinusoidal wave carrying the stress energy. Fracture is characterized as the final stage of deformation where the wave becomes solitary representing strain concentration and stops carrying the stress energy. Fracture always occurs along the strain concentration. The transitional behavior of the wave dynamics can be visualized as a change in the optical interferometric fringe pattern generated by the optical technique known as the Electronic Speckle-Pattern interferometry. Finite element analysis has been conducted to explain the experimentally observed behaviors and explore the mechanism of transition from to fracture.

**Keywords** Fracture dynamics · Deformation dynamics · Deformation wave · Optical interferometry

## Introduction

Theories of conventional fracture mechanics [1, 2] assume the existence of defects and discuss fractures based on the crack tip dynamics. The description is physically sound and applicable to variety of engineering cases. However, they do not connect fracture dynamics with deformation. This situation makes possible that mechanical failures are not foreseen at the time of a routine inspection because a visible defect does not exist and can lead to catastrophic incidences. A recent field theory [3, 4] has a mechanism to describe deformation and fracture on the same basis. This theory formulates deformation dynamics comprehensively by postulating that solids at any stage of deformation possess local elasticity and that the associated dynamics can be described by Hooke's law with the local coordinate system. Hooke's law is orientation preserving (translational) because the force and resultant deformation are mutually parallel. When deformation enters the plastic regime, the dynamics becomes nonlinear and hence the global coordinate system becomes unable to describe Hooke's law. The field theory solves this problem by introducing a connection field, which aligns all local frames in the same orientation so that Hooke's laws can be described at the global level. Thus, the connection field is rotational by nature. The overall dynamics can be described through physical law that governs the connection field. The interaction between the translational elasticity and the rotational connection field generates wave dynamics at the global level.

Previous experimental and theoretical studies [5, 6] indicate the propositions that plastic deformation is represented by sinusoidal waves that carry the stress energy, and that fracture is preceded by strain concentration represented by a solitary wave. However, the mechanism underlying the transition from deformation to fracture is still unclear. Understanding of this transition is of absolute importance in prediction of fracture and in related engineering. The aim of this study is to explore this transitional mechanism through numerical studies. Finite Element Models (FEM) are built to solve the wave equations derived from the field theory and results of numerical tests are discussed through comparison with experimental observations. It has been found that the transverse component of displacement (the component perpendicular to the applied load) plays an important role in transition from deformation to fracture.

---

S. Yoshida (✉) · D. R. Didie · S. Ashina
Department of Chemistry and Physics, Southeastern Louisiana University, Hammond, LA, USA
e-mail: syoshida@selu.edu

T. Sasaki · S. Takahashi
Department of Engineering, Niigata University, Niigata, Japan

## Theory

The wave characteristics of deformation fields can be derived from the field equations. Details of the field equations and their derivation can be found elsewhere [3, 4]. The field equations are in the following form.

$$\nabla \cdot \boldsymbol{v} = -j_0 \tag{7.1}$$

$$\nabla \times \boldsymbol{v} = \frac{\partial \boldsymbol{\omega}}{\partial t} \tag{7.2}$$

$$\nabla \times \boldsymbol{\omega} = -\frac{1}{c^2}\frac{\partial^2 \boldsymbol{\xi}}{\partial t^2} - \boldsymbol{J} \tag{7.3}$$

$$\nabla \cdot \boldsymbol{\omega} = 0 \tag{7.4}$$

Here $\boldsymbol{v}$ is the velocity of a point in the solid material, $\boldsymbol{\omega} = \nabla \times \boldsymbol{\xi}$ is the rotation, $j_0$ and $\boldsymbol{J}$ are the current of symmetry associated with the physical concept known as the local symmetry, and $c$ is the phase velocity of the shear wave [5].

$$c = \sqrt{\frac{G}{\rho}} \tag{7.5}$$

Here $G$ is the shear modulus and $\rho$ is the density. Substitution of Eq. (7.5) into Eq. (7.3) and rearrangement of terms yield the following equation.

$$\rho \frac{\partial^2 \boldsymbol{\xi}}{\partial t^2} = -G\nabla \times \boldsymbol{\omega} - G\boldsymbol{J} \tag{7.6}$$

On the right-hand side the first term represents the transverse force and the second term represents the longitudinal force acting on the unit volume represented by $\rho$. Integration of Eq. (7.2) with respect to time and use of the expression $\boldsymbol{\omega} = \nabla \times \boldsymbol{\xi}$ allow us to express the rotation vector in terms of the displacement vector and rewrite Eq. (7.6) as follows.

$$\rho \frac{\partial^2 \boldsymbol{\xi}}{\partial t^2} = G\nabla^2 \boldsymbol{\xi} - G\nabla (\nabla \cdot \boldsymbol{\xi}) - G\boldsymbol{J} \tag{7.7}$$

The longitudinal terms for the plastic and elastic dynamics are identified as follows (see p. 97 of [3]).

$$G\boldsymbol{J}_e = -(\lambda + 2G) \nabla (\nabla \cdot \boldsymbol{\xi}) \tag{7.8}$$

$$G\boldsymbol{J}_p = \sigma_0 \rho (\nabla \cdot \boldsymbol{v}) \boldsymbol{v} \tag{7.9}$$

Substitution of Eqs. (7.8) and (7.9) into (7.7) yields the following equation.

$$\rho \frac{\partial^2 \boldsymbol{\xi}}{\partial t^2} = G\nabla^2 \boldsymbol{\xi} - G\nabla (\nabla \cdot \boldsymbol{\xi}) + (\lambda + 2G) \nabla (\nabla \cdot \boldsymbol{\xi}) - \sigma_0 \rho (\nabla \cdot \boldsymbol{v}) \boldsymbol{v} \tag{7.10}$$

Noting that $\boldsymbol{v}$ is the temporal differentiation of $\boldsymbol{\xi}$, we can rearrange Eq. (7.10) as follows.

$$\frac{\partial^2 \boldsymbol{\xi}}{\partial t^2} + \sigma_0 (\nabla \cdot \boldsymbol{v}) \frac{\partial \boldsymbol{\xi}}{\partial t} - \frac{G}{\rho}\nabla^2 \boldsymbol{\xi} = -\frac{G}{\rho}\nabla (\nabla \cdot \boldsymbol{\xi}) + \frac{(\lambda + 2G)}{\rho}\nabla (\nabla \cdot \boldsymbol{\xi}) \tag{7.11}$$

Equation (7.11) is a decaying wave equation with the source term (the right-hand side). The second term on the left-hand side originates from the plastic longitudinal force (7.9), the third term on the left-hand side and the first term on the right-hand side originate from the transverse force in Eq. (7.6), and the second term on the right-hand side represents the elastic dynamics.

## Numerical Analysis

### Finite Element Model (FEM)

We built a FEM as a two-dimensional model using an $xy$-coordinate system in the coefficient form as follows.

$$\frac{\partial^2 \xi_x}{\partial t^2} + d_a \left(\nabla \cdot \boldsymbol{v}\right) \frac{\partial \xi_x}{\partial t} - \frac{G}{\rho} \frac{\partial^2 \xi_x}{\partial x^2} - \frac{G}{\rho} \frac{\partial^2 \xi_x}{\partial y^2} = -\frac{G}{\rho} \frac{\partial}{\partial x} \left( \frac{\partial \xi_x}{\partial x} + \frac{\partial \xi_y}{\partial y} \right) + \frac{(\lambda + 2G)}{\rho} \frac{\partial}{\partial x} \left( \frac{\partial \xi_x}{\partial x} + \frac{\partial \xi_y}{\partial y} \right) \quad (7.12)$$

$$\frac{\partial^2 \xi_y}{\partial t^2} + d_a \left(\nabla \cdot \boldsymbol{v}\right) \frac{\partial \xi_y}{\partial t} - \frac{G}{\rho} \frac{\partial^2 \xi_y}{\partial x^2} - \frac{G}{\rho} \frac{\partial^2 \xi_y}{\partial y^2} = -\frac{G}{\rho} \frac{\partial}{\partial y} \left( \frac{\partial \xi_x}{\partial x} + \frac{\partial \xi_y}{\partial y} \right) + \frac{(\lambda + 2G)}{\rho} \frac{\partial}{\partial y} \left( \frac{\partial \xi_x}{\partial x} + \frac{\partial \xi_y}{\partial y} \right) \quad (7.13)$$

Relating the normal strains along the $x$-axis and $y$-axis with Poisson ratio $\nu$ and introducing parameter $\alpha$ to express the degree of the elastic dynamics, we can rewrite Eqs. (7.12) and (7.13) as the following differential equations.

$$\frac{\partial^2 \xi_x}{\partial t^2} + d_a \left(\nabla \cdot \boldsymbol{v}\right) \frac{\partial \xi_x}{\partial t} - \frac{G}{\rho} \frac{\partial^2 \xi_x}{\partial y^2} = \frac{\nu G}{\rho} \frac{\partial^2 \xi_x}{\partial x^2} + \alpha \frac{(\lambda + 2G)(1-\nu)}{\rho} \frac{\partial^2 \xi_x}{\partial x^2} \quad (7.14)$$

$$\frac{\partial^2 \xi_y}{\partial t^2} + d_a \left(\nabla \cdot \boldsymbol{v}\right) \frac{\partial \xi_y}{\partial t} - \frac{G}{\rho} \frac{\partial^2 \xi_y}{\partial x^2} = -\frac{G}{\rho} \frac{\partial^2 \xi_x}{\partial x \partial y} + \alpha \frac{(\lambda + 2G)}{\rho} \frac{\partial}{\partial y} \left( \frac{\partial \xi_x}{\partial x} + \frac{\partial \xi_y}{\partial y} \right) \quad (7.15)$$

The FEM solves Eqs. (7.14) and (7.15) using $\lambda = 2\nu G/(1 - 2\nu)$ [7, 8].

Figure 7.1 illustrates the geometry used for the FEM. The specimen is a two-dimensional rectangular plate. The origin of the $(x, y)$ coordinate is at the bottom left corner of the specimen. A monotonically increasing or sinusoidal displacement in the form of $A \sin \omega t$ is given to the right side at $x = 2$ (m) as the boundary condition. Here $\omega = 2\pi f$ is the driving angular frequency. Point $P(0.5, 0.25)$ is an arbitrary set reference point where the temporal behavior of the displacement is analyzed.

### General Argument Regarding Transition to Fracture

Experiments based on Electronic Speckle-Pattern Interferometry (ESPI) conducted previously [5, 6, 9, 10] indicate that fracture of plate specimen is always preceded by the formation of strain concentration. This type of strain concentration is observable as a concentrated fringe patterns consisting of mutually parallel linear dark lines. The timing and detailed characteristics regarding the formation of the strain concentration depends on the type of the material and loading condition, tensile [5, 6] or cyclic loading [10], but it always appears prior to the final fracture. Here the fringe pattern is formed with the

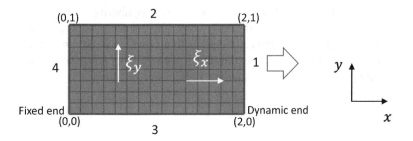

**Fig. 7.1** Finite element model (FEM) geometry

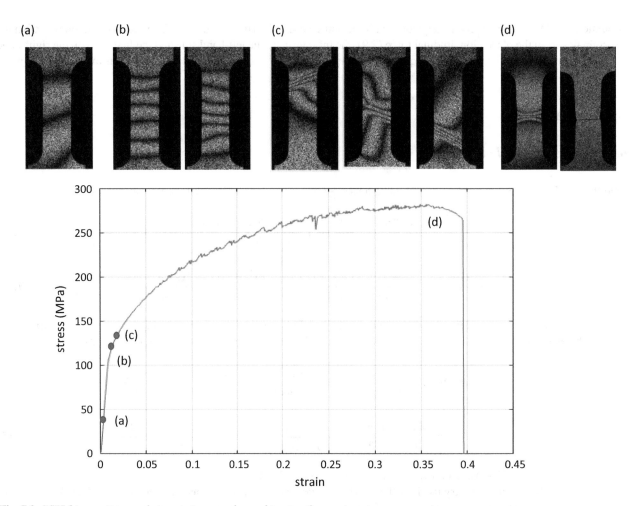

**Fig. 7.2** ESPI fringe patterns and stress train curve observed in a tensile experiment

use of the ESPI technique known as the subtraction method (see Chap. 7 of [3]); the specimen being loaded by a test machine is illuminated by a pair of laser beams from an optical interferometric setup and its image is captured by a digital camera continuously with a constant time step. Each of these images contains the relative phase difference between the two laser beams at all points of the specimen and called an interferogram. By electronically subtracting the interferogram obtained at a time step from that obtained at another time step, one can form the so-called interferometric fringe pattern. The images shown in Fig. 7.2 are typical examples of fringe pattern. The dark lines observed in a fringe pattern represent the contour of the differential displacement that the specimen undergoes between the two time-steps used for the fringe pattern formation.

Figure 7.2 presents fringe patterns obtained in a tensile experiment with a constant pulling rate conducted on an aluminum alloy (A5053) plate (gauge length 25 mm, width 10 mm, thickness 3 mm) specimen along with the stress-strain curve. Labels (a)–(d) indicated in the stress-strain curve show when the fringe patterns with the respective labels are formed. Image (a) represents a typical fringe pattern observed before the yield point. The dark fringes are approximately equidistant and parallel to one another. Images labeled (b) are typically observed immediately after the yield point. The number of dark fringes is significantly greater than image (a) and some of them are closer to the neighboring one than the others. Images labeled (c) show the strain concentration mentioned above. In the early stage, the concentrated strain moves across the specimen as the three images in (c) indicate. Image (d) is a typical fringe pattern observed at the final stage of deformation where the concentrated strain is stationary. Fracture occurs along the line of the stationary strain concentration. The right-most image in Fig. 7.2 shows that the fracture line runs along the stationary strain concentration.

## Numerical Tests

The FEM was used under several conditions listed in Table 7.1.

**Realistic test** The aim of this numerical test is to see if the FEM reproduces the type of fringe patterns observed experimentally. The transverse wave velocity and the pulling rate shown in Table 7.1 are of the same order as experimental values [5]. Figure 7.3 shows the two-dimensional profile of the velocity components parallel to the tensile axis (the longitudinal component $\dot{\xi}_x$) and perpendicular to it (transverse component $\dot{\xi}_y$), respectively. Since the experimental fringe pattern represents the differential displacement that the specimen undergoes during the interval between the two time-steps, the contour is proportional to the velocity. It is legitimate to compare the velocity field profiles with experimental fringe patterns.

Figure 7.3a, b are obtained when the average longitudinal strain is 17% and 38%, respectively. Notice that at stage (a) the longitudinal velocity is an order of magnitude higher than the transverse velocity but at stage (b) the transverse velocity is factor of two higher than the longitudinal velocity. This indicates that the transverse velocity grows faster than the longitudinal velocity.

There are similarities in the general pattern between the numerical velocity field profiles and experimental fringe patterns. Figure 7.4 presents ESPI fringe patterns observed in a tensile experiment on an aluminum alloy A7075 (pulling rate 0.017 mm/s) along with the loading curve [6]. (a)–(g) indicate the stage of deformation when each pair of fringe pattern is recorded. Pay attention to the transverse ($\Delta\xi_y$) patterns at stage (d) and (e). The former resembles the numerical transverse profile (a) in Fig. 7.3 in that there are two large circular patterns near the two sides of the specimen. The latter is similar with the numerical transverse profile (b) in Fig. 7.3 in that the pattern appears to be diagonal across the width of the specimen near

**Table 7.1** Numerical tests conditions

| Run ID | G (N/m$^2$) | $\nu$ | $\rho$ (kg/m$^3$) | $v_L$ (mm/s) | $v_T$ (mm/s) | $V_{pull}$ (mm/s) | | | $F_{driv}$ (mHz) | | |
|---|---|---|---|---|---|---|---|---|---|---|---|
| Realistic test | $2.2 \times 10^{-4}$ | 0.4 | 1.0 | 9.4 | 14.8 | 0.45 | | | – | | |
| High Poisson | $1 \times 10^{-3}$ | 0.4 | 1.0 | 20 | 31.6 | 1.4 | 0.62 | 0.45 | 500 | 50 | 0.5 |
| Low Poisson | $1 \times 10^{-3}$ | 0.1 | 1.0 | 10 | 31.6 | 1.4 | 0.62 | 0.45 | 500 | 50 | 0.5 |

$v_L/v_T$: longitudinal/transverse wave velocity; $V_{pull}$: pulling rate; $F_{driv}$: driving frequency

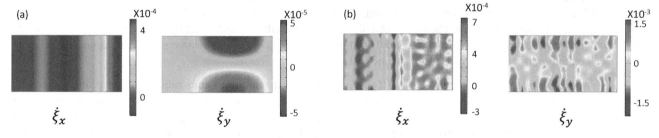

**Fig. 7.3** Two-dimensional profiles of velocity field from numerical test 1; (**a**) average strain 17%, (**b**) average strain 38%

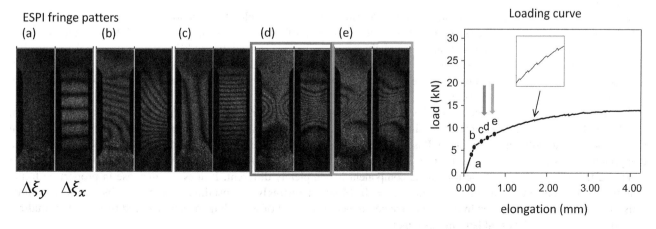

**Fig. 7.4** Experimental fringe patterns in longitudinal and transverse differential displacement [6]

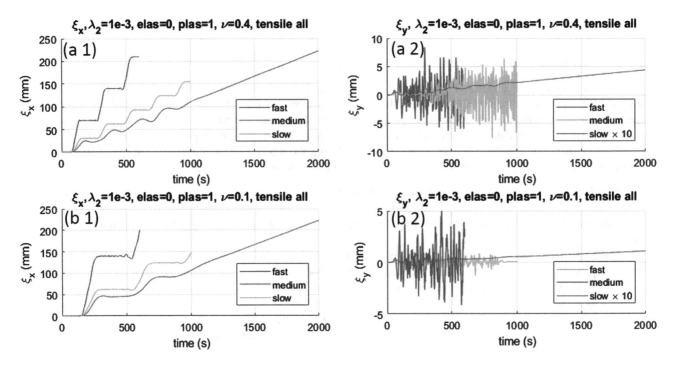

**Fig. 7.5** Numerical test for tensile loading under different pulling rates and Poisson's ratio

the central part. The longitudinal patterns are also similar between the numerical and experimental. At the earlier stage the contours are perpendicular to the tensile axis whereas at the later stage the longitudinal pattern has some element similar with the transverse pattern. It is naturally interpreted that the central diagonal pattern observed in the transverse image develops to the strain concentration seen in Fig. 7.1c.

**Effects of pulling rate and Poisson ratio** Figure 7.5 shows temporal variation of $\xi_x$ and $\xi_y$ at point $(0.5, 0.25)$ resulting from numerical tests "High Poisson" and "Low Poisson" conducted under the tensile load with three different pulling rates. Here $\xi_x$ and $\xi_y$ are the longitudinal and transverse total displacement (not differential), pulling rates labeled "fast", "medium" and "slow" represent pulling rates of 1.4, 0.62 and 0.45 mm/s (Table 7.1). The elastic force term is set to null for all these tests ($\alpha = 0$ on the right-hand sides of Eqs. (7.14) and (7.15)).

First, consider the effect of pulling rate. (a1) and (b1) indicate that for both Poisson's ratio the faster pulling rate tends to generate oscillatory behavior more greatly in the longitudinal component. This is easily understood by imagining pulling one end of a spring faster and slower. If pulling faster, the stretching behavior of the spring is localized near the pulling end because the stretching pattern (compression wave) needs to travel along the spring at a certain wave velocity. On the other hand, if the same end is pulled more slowly, the compression wave has time to travel to the other end. Consequently, the entire spring tends to move together.

In the above dynamics, the oscillation period is determined by the phase velocity. Equation (7.14) indicates that the phase velocity of the longitudinal wave is proportional to the square root of the product of the elastic modulus and the Poisson's ratio. The three plots in (a1) have the same periodicity because the elastic modulus and Poisson's ratio are the same. The periodicity in (b1) is longer than (a1) because the Poisson's ratio is lower, hence the wave velocity is lower. The wavelength is the same for (a1) and (b1) because it is determined by the geometry. Consequently, the faster wave velocity case results in the higher frequency or shorter period.

It is extremely interesting to look at the behavior of the transverse component in (a2) and (b2). In both cases in common, the slow pulling case does not show an oscillatory behavior. Instead, the transverse component increases monotonically in proportion to the longitudinal component. It is obvious that this behavior represents the Poisson's effect. In fact, the transverse component increases more in (b1) than (b2) as the Poisson's ratio is higher for the former. When pulled faster, on the other hand, the temporal behavior of the transverse component is completely different. It does not increase in proportion to the longitudinal component but oscillatory around $\xi_y = 0$. More interestingly, the oscillation amplitude increases with time. This indicates that the oscillation in the transverse component is a result of wave dynamics transferred from the longitudinal component as will be discussed later in this paper.

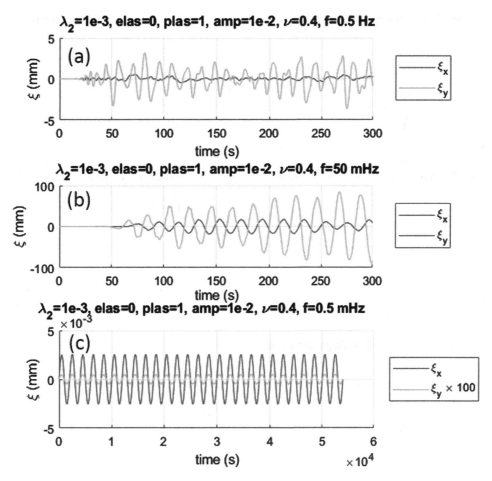

**Fig. 7.6** Numerical test under cyclic loading with three driving frequencies

Figure 7.6 is the cyclic loading version of Fig. 7.5 with a common driving amplitude (10 mm) and three different driving frequencies; (a) 500 mHz, (b) 50 mHz and (c) 0.5 mHz. The three cases show significantly different behaviors from one another. The oscillation amplitude for (b) is orders of magnitude greater than (a) and (c). It is possible to discuss this observation based on the resonance. The resonant frequency for the longitudinal oscillation can be evaluated as $f_r = (2N - 1)v_p/(2L)$ where $N$ is the integer representing the mode, $v_p$ is the phase velocity of the wave and $L$ is the specimen length. From the wave velocity shown in Table 7.1, the resonant frequency can be estimated as 5 mHz. The driving frequency of 500 mHz is too high (too blue side of the spectrum) as compared with the resonant frequency, and hence the amplitude of the longitudinal wave is low. The driving frequency of 50 mHz is close enough to the resonant frequency and therefore the amplitude of the longitudinal oscillation is high. The periodicity observed in (b), five oscillation in 100 s, is the driving frequency of 50 mHz. Note that while the amplitude of the longitudinal oscillation is constant, the amplitude of the transverse oscillation grows with time, as is the case of tensile load observed in Fig. 7.5.

The lowest driving frequency (c) does not indicate a wave-like behavior. Both the longitudinal and transverse displacements oscillate at the driving frequency with a constant amplitude. This case corresponds to the lowest pulling rate observed in Fig. 7.5. The loading motion is too slow to excite waves in the specimen.

Figure 7.7 presents the effect of Poisson's ratio on the longitudinal and transverse oscillatory behaviors under the cyclic loading at 50 mHz. With the lower Poisson's ratio of 0.1, the longitudinal displacement does not show a clear oscillatory behavior even at the driving frequency. This is understandable because on the right-hand side of Eq. (7.14) the lower the Poisson's ratio the less the longitudinal dynamics wave-like. A very interesting observation in Fig. 7.7 is the behavior of the transversal displacement. Although the amplitude is a factor of two smaller than the higher Poisson's ratio case, (b2) clearly shows the oscillatory feature at the driving frequency and, similar with (a2), the oscillation amplitude increases with time.

**Effects of elastic force** Figure 7.8 compares the behaviors of longitudinal and transverse displacements for "High Poisson" case with the fast pulling and 500 mHz cyclic loading without the elastic force terms (a1, b1) and with (a2, b2) the elastic

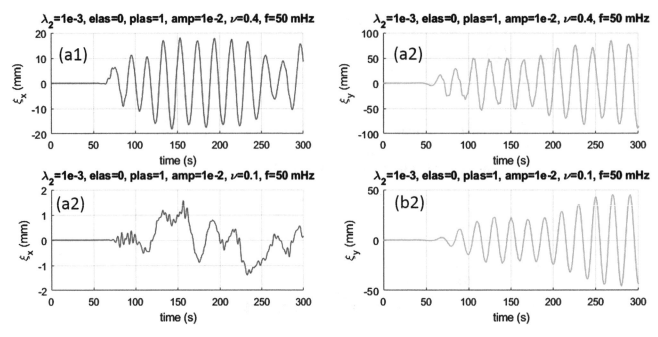

**Fig. 7.7** Effect of Poisson's ratio under cyclic loading with 50 mHz driving frequency

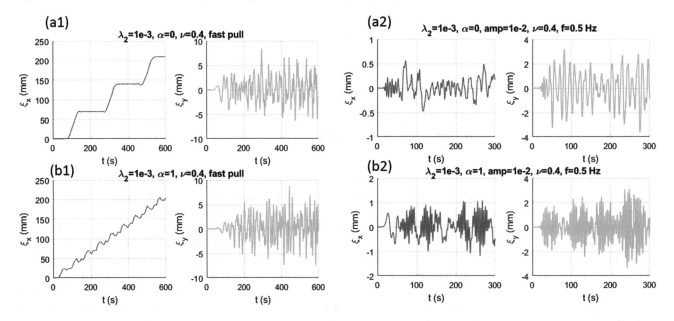

**Fig. 7.8** Effect of elastic force term

force term. The elastic force term increases the oscillation frequency of $\xi_x$ as expected. On the right-hand side of Eq. (7.14), the inclusion of the elastic force term results in the increase in the wave velocity.

## Concluding Remarks

This study yields the following findings.

1. The loading speed affects the generation of wave dynamics significantly. Faster loading tends to generate the wave dynamics more easily. When loaded slower than a certain level, the wave dynamics is not generated.

2. The Poisson's effect and the elastic modulus change the frequency of the longitudinal wave in proportion to the square root of the product.
3. The elastic force term affects the longitudinal wave dynamics but has little effect on the transverse wave dynamics.
4. In the case of cyclic loading, the generation of wave dynamics is much greater if the driving frequency is close to the resonant frequency of the longitudinal vibration.

The observations made in this study leads to the following proposition. Once the wave dynamics is excited in the longitudinal component by the load, the wave dynamics is transferred to the transverse component via $G \nabla \times \omega$ term of the equation of motion (7.6). In this process, the longitudinal dynamics is controlled by the external agent through the boundary where the load is applied. However, there is no mechanism to control the transverse dynamics. The longitudinal dynamics keeps transferring the oscillation energy to the transverse dynamics. Therefore, with the passage of time, the amplitude of the transverse oscillation increases. It is likely that when the amplitude of the transverse displacement reaches the critical value, a shear band is formed. Subsequently the fracture criterion is satisfied, and the specimen breaks along the shear band. This proposition is yet to be verified by experiment.

# References

1. Griffith, A. A. (1920). *Philosophical Transactions of the Royal Society A, 221*:, 163–198.
2. Irwin, G. R. (1948). Fracture dynamics. In *Fracturing of metals*. Cleveland: American Society for Metals.
3. Yoshida, S. (2015). *Deformation and fracture of solid-state materials*. New York: Springer.
4. Yoshida, S. (2015). Comprehensive description of deformation of solids as wave dynamics. *Mathematics and Mechanics of Complex Systems, 3*(3), 243–272.
5. Yoshida, S., Siahaan, B., Pardede, M. H., Sijabat, N., Simangunsong, H., Simbolon, T., & Kusnowo, A. (1999). Observation of plastic deformation wave in a tensile-loaded aluminum-alloy. *Physics Letters A, 251*, 54–60.
6. Sasaki, T., Suzuki, H.& Yoshida, S. Evaluation of dynamics deformation behavior of aluminum alloy by electronic speckle pattern interferometry. In: *Imaging Methods for Novel Materials and Challenging Applications*, 3, *Conference Proceedings of Society for Experimental Mechanics Series* (pp. 133–139.). New York: Springer.
7. Spencer, A. J. M. (1980). *Continuum mechanics*. London: Longman.
8. Marsden, J. E., & Hughes, T. J. R. (1983). *Mathematical foundation of elasticity*. Englewood Cliffs: Prentice-Hall.
9. Yoshida, S., Muchiar, Muhamad, I., Widiastuti, R., & Kusnowo, A. Optical interferometric technique for deformation analysis. *Optics Express, 2*, 516–530. (focused issue on "Material testing using optical techniques").
10. Yoshida, S., Ishii, H., Ichinose, K., Gomi, K., & Taniuchi, K. (2004). Observation of optical interferometric band structure representing plastic deformation front under cyclic loading. *Japanese Journal of Applied Physics, 43*, 5451–5454.

# Chapter 8
# Fatigue Monitoring of a Dented Pipeline Specimen Using Infrared Thermography, DIC and Fiber Optic Strain Gages

J. L. F. Freire, V. E. L. Paiva, G. L. G. Gonzáles, R. D. Vieira, J. L. C. Diniz, A. S. Ribeiro, and A. L. F. S. Almeida

**Abstract** An investigation program has been launched with the objective of presenting combinations of analytical, experimental and numerical methods to predict and monitor fatigue initiation and fatigue damage progression in equipment such as pressure vessels, tanks, piping and pipelines with dents or shape anomalies. The present paper reports initial results from tests where these techniques were applied to a pipeline specimen containing a plain longitudinal dented subjected to hydrostatic cyclic loading. Some of the material's fatigue properties assessment used validated rapid approaches based on infrared thermography. The monitoring of fatigue initiation and propagation in the actual specimen used nondestructive infrared inspection techniques. Thermoelasticity stress analysis (TSA) and three-dimensional digital image correlation (3D-DIC) were used to determine fatigue hot spots locations as well as strain concentrations. Full field TSA and fiber optic Bragg strain gages (FBSG) were used to determine the overall stress field (TSA) as well hot spot strain evolution (FBSG) along the loading cycles. Strain fields determined from the experimental measurements and from finite element analysis (FEA) were combined with the fatigue Coffin-Manson model to predict fatigue life (Nc). The tested 3 m long tubular specimen was fabricated with API 5L Gr. B 12.75″ OD with ¼″ thickness pipes. The excellent agreement among test and predicted results achieved up to now are commented in the paper.

**Keywords** Fatigue · Thermoelasticity · Digital image correlation · Pipeline · Dent

## Introduction

Pipelines represent the main and safe economic alternative to transport liquid or gaseous products. Among the different types of pipeline damage, mechanical defects such as dents are very dangerous and hazardous for society, economy and the welfare of the structures. Therefore, the study of their structural integrity deserves special attention, e.g. [1–8], since their failure and loss of containment can cause safety and environmental disasters as well as operational setbacks. Thus, predicting, monitoring and detecting fatigue damage in pipelines and by extension to piping and pressure vessels with dents have been of paramount importance.

This paper uses data acquired during fatigue tests of a real scale dented pipeline specimen. A 3 m long tubular specimen with a plain longitudinal dent was fabricated with pipe grade API 5L Gr. B and had a 12.75″ (323 mm) outside diameter (OD) with ¼″ (6.35 mm) thickness (t). Material properties were determined using ASTM standards and also fatigue rapid assessment methods proposed by Risitano and co-workers [9–12]. Numerical and experimental techniques were applied to monitor and assess fatigue damage. During the testing phase, experimental data was acquired using Thermoelasticity (TSA) [7, 8, 13–19], Digital Image Correlation (DIC) [7, 8, 20, 21] and Fiber Optic Bragg Strain Gages (FBSG) techniques [8]. In the analysis phase strain fields determined from the experimental measurements and from finite element analysis (FEA) were combined with the fatigue Coffin-Manson strain-life equation to predict fatigue life (Nc) [22].

Excellent agreement between the predicted and measured results was achieved, showing the advantages of coupling different techniques in order to not only to assess crack initiation and monitor fatigue development, but also to show their applicability to real size structures.

---

J. L. F. Freire (✉) · V. E. L. Paiva · G. L. G. Gonzáles · R. D. Vieira · J. L. C. Diniz · A. S. Ribeiro
Pontifical Catholic University of Rio de Janeiro, PUC-Rio, Rio de Janeiro, RJ, Brazil
e-mail: jlfreire@puc-rio.br; gonzalesglg@aaa.puc-rio.br; diniz@puc-rio.br; alexsr@puc-rio.br

A. L. F. S. Almeida
Petrobrás SA/CENPES, Cidade Universitária, Rio de Janeiro, RJ, Brazil
e-mail: anafampa@petrobras.com.br

## Material and Experimental Procedure

Uniaxial test specimens from the pipe material, Fig. 8.1a, were tested in a 100 kN INSTRON servo-hydraulic machine under static monotonic load and cyclic axial load. Relevant material data are given in Table 8.1 and Fig. 8.2. The uniaxial test specimens were machined from longitudinal strips cut from the thin walled tested pipe specimen. The dented pipe specimen (Fig. 8.1b) was subjected to cyclic internal pressure by means of a water loading device coupled to a 500 kN MTS servo-hydraulic machine.

During each static or fatigue test, the surface temperature of the specimens was recorded in real time by a thermocamera FLIR A655sc (640 × 480 uncooled micro-bolometers, 50 Hz acquisition rate, 17 μm spatial resolution, 30 mK sensitivity). All uniaxial and pipe specimens were painted with a thin layer of black paint to increase emissivity. Besides the black paint, the dented pipe specimen had also painted white dots spread over the black paint as depicted in Fig. 8.1b. The small white dots were needed by the digital image analysis technique used to determine surface strains. The temperature data was acquired and analyzed using the ResearchIR software from FLIR. TSA data was acquired with DeltaTherm II software from Stress Photonics.

The pipe specimen was initially deformed in a 500 kN press to reach a dent depth of 48% of the external diameter. Plane head caps were welded to close the pipe specimen that was filled with water and pressurized to 6.5 MPa, this process causing re-rounding of the dent forming its plain shape with a final maximum depth of 15.74% of the pipe external diameter. The final dent shape was fully mapped with caliper measurements and also by using 3D-DIC. The longitudinal profile of the re-rounded dent is depicted in Fig. 8.3. Caliper and DIC measurements agreed with deviation of less than 1%.

The DIC technique employed the software VIC-Snap and VIC-3D from Correlated Solutions Inc. respectively for image acquisition and image analysis. The stereoscopic system consisted of two 5MP cameras (Point Grey GRAS-50S5M) including high magnification lenses (Tamron AF28-200 mm F/3.8–5.6). A first pair of images was taken, after re-rounding, while the

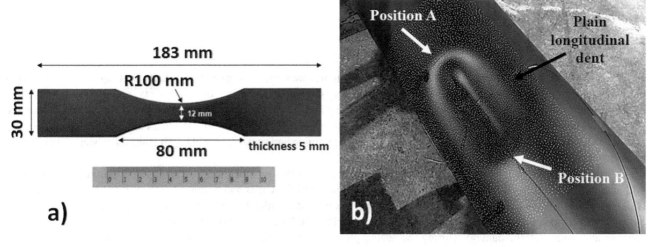

**Fig. 8.1** (a) Uniaxial specimen's geometry; (b) Dented pipe specimen black painted for the thermographic analysis and with small white dots to allow for digital image correlation analysis

**Table 8.1** Material properties (API 5L Gr. B) [23, 24]

| Actual chemical composition | 0.13C, 0.54Mn, 0,01Si and low P, S |
|---|---|
| Specified minimum yield strength, SMYS [24] | 245 MPa |
| Specified minimum ultimate strength, SMUS [24] | 415 MPa |
| Yield Strength Sy (average of 3 specimens, 0.5% total strain) | 316 MPa |
| Engineering ultimate strength Su (average of 3 specimens) | 420 MPa |
| True ultimate strength Su (average of 3 specimens) | 500 MPa |
| Maximum strain at fracture $\varepsilon_f$ | 0.36 |
| Fatigue limit Se R = −1 (average of 7 specimens) | 253 MPa |
| Fatigue limit Se (average of 3 specimens) R = 0.025 | 183 MPa |
| Fatigue limit Se (average of 3 specimens) R = 0.1 | 157 MPa |
| Young modulus (average of 3 specimens), E | 182 GPa |

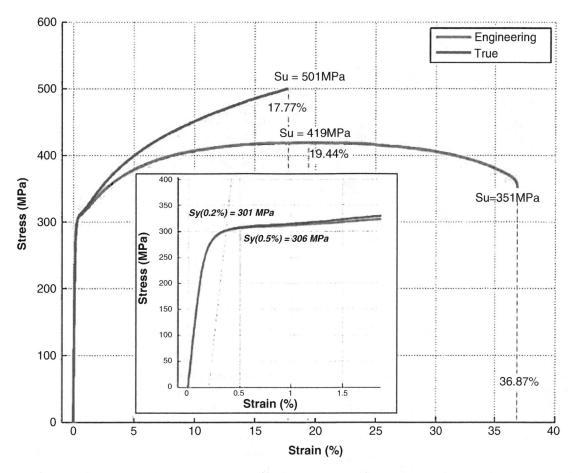

**Fig. 8.2** Tensile stress-strain curve of one of the three specimens of API 5L Gr. B pipe [23]

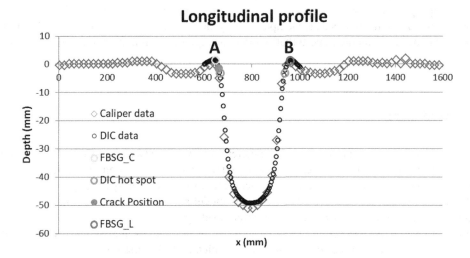

**Fig. 8.3** Longitudinal dent profile after re-rounding from caliper and DIC measurements. Circumferential and longitudinal FBSG and DIC hot spots at positions A and B

pressure was zero. After, sequences of images were made while the pipe internal pressure was increased in steps of 0.05 MPa until reaching the maximum pressure (6.0 MPa). This process was repeated for the first two initial pressuring cycles. DIC data analysis used the correlation algorithm NSSD (normalized sum of squared differences), a subset size of 35 × 35 pixels, a grid step of 8 pixels, a strain window of 15 × 15 displacement points and a pixel size about 200 μm. An example of the full field dent shape determined using the 3D-DIC technique is given in Fig. 8.4.

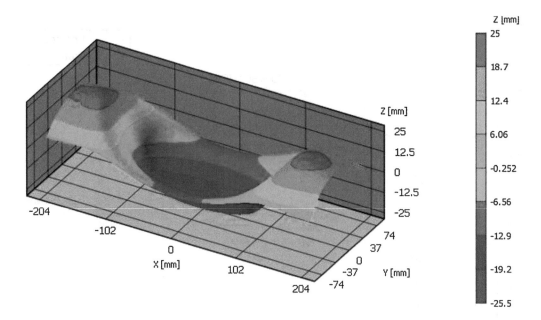

**Fig. 8.4** Dent geometry after re-rounding determined from 3D-DIC measurements

The optical fiber (FBGS©) used in the strain measurements contained two Bragg grating with gage length size of 8 mm, each grating having distinct wavelengths (1514 and 1524 nm), which allowed multiplexing the strain signal. The optical fiber employed had 20% reflectivity and a 0.1 nm spectral bandwidth with an Ormocer coating.

The dented thin walled pipe specimen was subjected to cyclic internal pressure by means of a water loading device coupled to a 500 kN MTS servo-hydraulic machine. Test frequency was 1.0 Hz. Cyclic pressure ranges were applied to the pipe specimen in two sequences. The first sequence was composed of 2 cycles of pressure ranging from 0 to 6.0 MPa. This first sequence was used to define the relationship of strain distribution (DIC acquired) and applied pressure. After, the second loading sequence was applied with pressure ranging from 0.2 to 6.2 MPa. This loading sequence was maintained until pressurized water leakage occurred from the most critical hot-spot (crack position). During this loading sequence, the TSA technique was applied to monitor the dent surface temperature cyclic variation as well as (while valid) its stress distribution ($\Delta\sigma_1 + \Delta\sigma_2$), allowing to detect temperature perturbations due to fatigue crack initiation and due to crack propagation across the pipe wall thickness.

## Finite Element Model (FEM)

Two finite element models were constructed using a commercial software package (ANSYS v15.0) to simulate the dented specimen under internal pressure. The true-stress-true strain curve of Fig. 8.2 was employed in the first model. The first model used ¼ of symmetry with restrictions to the displacements in the faces of symmetry, consisting of 3923 elements (SOLID186-CONTA174-TARGE170) and 26,885 nodes. This model simulated the entire initial indentation and the later re-rounding processes, as well as the first two cycles of pressure loading.

Figure 8.5 presents the variation of the circumferential strain of the hot-spot located at point A and its symmetrical point, B. The circumferential strain varies during the indentation as depicted by the first time-step defined from 0 to 1 s. Punch unloading occurs along time-steps defined from 1 to 4 s. Re-rounding of the initial dent occurs due to the application of internal pressure to a maximum of 6.5 MPa (time-steps defined from 4 to 12 s). Depressurization occurs at time-steps defined from 12 to 13 s. A dent shape similar to the one presented in Figs. 8.3 and 8.4 results after decreasing the internal pressure to zero (time 13 s). Finally, two cycles of internal pressure ranging from 0 to 6.0 MPa (time steps defined from 13 to 17 s) were applied. The analysis of the time steps defined from 13 to 17 s (Fig. 8.5) shows that the variation of the circumferential strain after of the execution of first cycle (0–6.0 MPa) corresponds to a linear elastic behavior. Therefore, the second FE model was developed using the elastic part of the stress-strain curve of Fig. 8.2. The second finite element analysis combined the "exact" dent shape measured by the DIC technique (Fig. 8.4) with the quasi-cylindrical portion of the non-deformed pipe to build the new geometric model as depicted in Fig. 8.6.

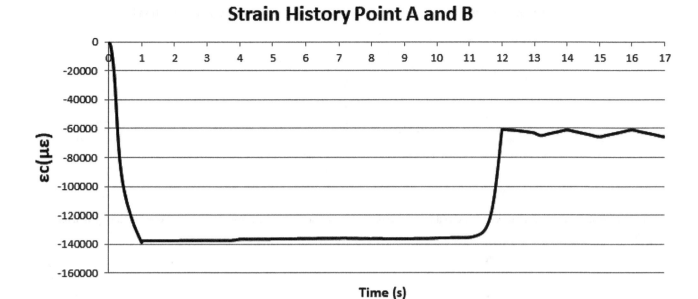

**Fig. 8.5** Circumferential strain history of point A and B (Fig. 8.3) along the indentation process, re-rounding and application of two cycles of pressure (0–6.0 MPa)

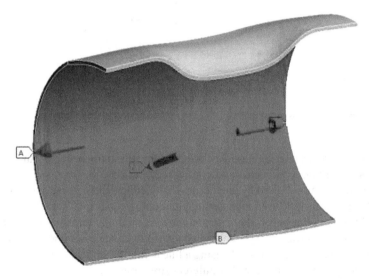

**Fig. 8.6** Finite element model of the dent and surrounding cylindrical pipe, constructed from the "exact" shape of the dent determined by the DIC technique. Model used 88,701 elements and 449,184 nodes, SOLID186, CONTA174 and TARGE170

## Results and Discussion

It was found a good agreement between and strain results measured by the DIC and FBSG techniques at positions located at points A and B as well as with results determined from the second FE model for those points. Figure 8.7 compares the circumferential strain measured by DIC technique (last cycle of the first load sequence) with the linear elastic finite element results determined for the same position.

Figure 8.8 shows the strains measured by the Bragg strain gages along the first load sequence compared with DIC and FE results (the FE results were translated along the abscissa in order to match the pressurization cycle to easy the comparison of strain plots).

Taking into consideration the meridional line that passes through the center of the dent, as shown in Fig. 8.3, DIC, TSA and FE results were assessed along it. These results are plotted in Fig. 8.9, in terms of the sum of the principal stress for a pressure range equal to 6.0 MPa, used in the second load sequence. Some discrepancy is expected among the TSA, the DIC

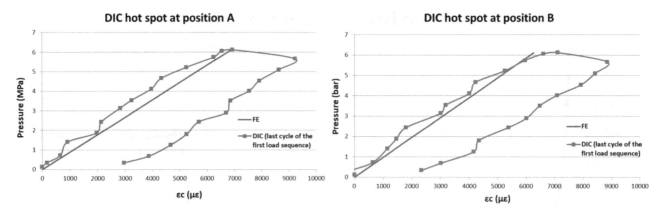

**Fig. 8.7** Circumferential strain measured by DIC at the last cycle of the first load sequence and calculated by the FEM at DIC hot spots at positions A and B

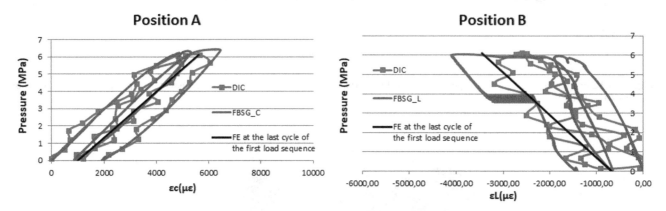

**Fig. 8.8** Circumferential and longitudinal strains measured by FBSG and DIC and calculated by FE at the positions where the FBSG were placed

and FE results. While TSA is in essence a linear elastic technique, some plasticity occurs in and around the hot spot positions even if a double flow stress magnitude is expected and considered. Besides, the TSA signal is composed of a magnitude proportional to the temperature variation caused by the thermoelastic effect, and a phase angle, which is related to the signal of the actuating stresses. Vieira et al. [25] used a phase angle modulus equal to 45° to differentiate the magnitude signal (tension or compression). In the present paper, the best phase angle modulus that showed good correspondence with the results of the other applied techniques was 20°. The relation between the TSA magnitude signal (cam units provided by the camera) and the principal stress state (MPa) was made using a linear calibration relationship established by Paiva et al. [26], where low carbon steel tensile specimens were tested at different stress ranges.

Figure 8.10 shows the plot of the TSA magnitude (measured in camera units), the phase angle and the first invariant stress response along the meridional line of the dent. Some noise in the signal may have origin in the pattern of white dots applied on the surface of the specimen applied to facilitate the use of the DIC technique.

The TSA output (proportional to the principal stress invariant range) is presented in Fig. 8.11. The TSA image was taken when the pressure range was 6.0 MPa and it was gathered after 1000 pressurizing cycles of the second load sequence. It has to be noted that for some locations, at this pressure range, some plasticity is occurring, and therefore the TSA results obtained at these locations must be interpreted as qualitative results. This Fig. 8.11 shows the critical hot-spot at the edges of the dent as well as (caused by different temperatures) the uniaxial fiber optic strain gage locations, one of them measuring circumferential strain and another measuring longitudinal strain at reasonably symmetrical points. The two circumferential FBSG at the center of the dent were not considered in the present analysis since the stress states at this position are irrelevant when compared to the hot spot positions.

TSA results collected along the cyclic hydrostatic load sequence and carried out with the dented thin walled pipe specimen presented in Fig. 8.12. The TSA images show the crack initially developing at 5600 cycles and after it has propagated through the specimen wall thickness at 6500 cycles.

Fatigue life (Nc) of the hot spot location was calculated using the Coffin-Manson equation (8.1). It used the universal fatigue exponents 0.12 and 0.6 proposed by Manson [22] and data from Table 8.1. Calculations did not took into consideration

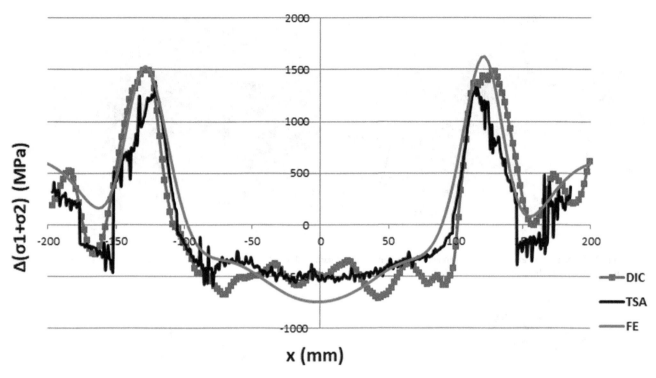

**Fig. 8.9** TSA, DIC and FE results plotted for points located at the meridional symmetric line of the dent

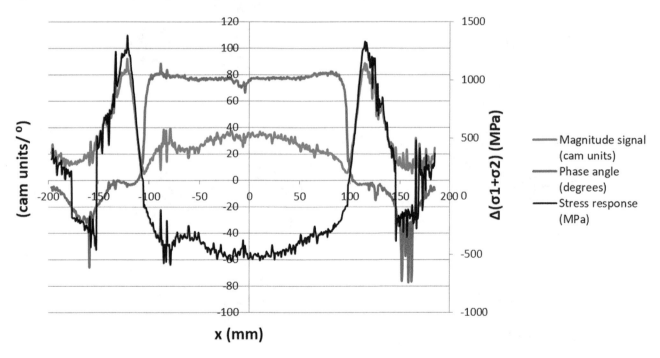

**Fig. 8.10** TSA magnitude, phase signal and stress response along the meridional symmetric line of the dent

the mean cycle stress ($\sigma_m = 0$). One reason for this was the fact that, although the mean stress cycle caused by the actuating pressure was known, the total mean stress is unknown due to the uncertain previous load history imposed to the pipe material. Part of this history can be inferred from the indentation process using the first FEM and pressure-stress-strain results such as presented in Fig. 8.5, but the previous history caused by the pipe fabrication process is difficult to be accurately simulated.

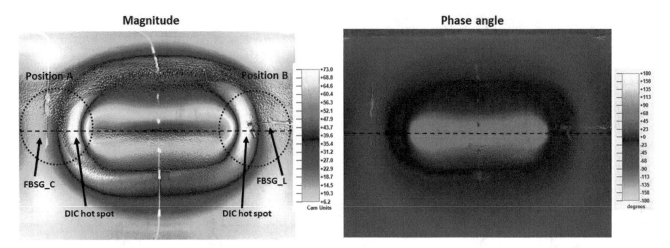

**Fig. 8.11** TSA magnitude and phase angle image of the dented area taken under 6.0 MPa internal pressure range variation (1 Hz). The critical point highlighted in the figure coincides with hot spots at positions A and B

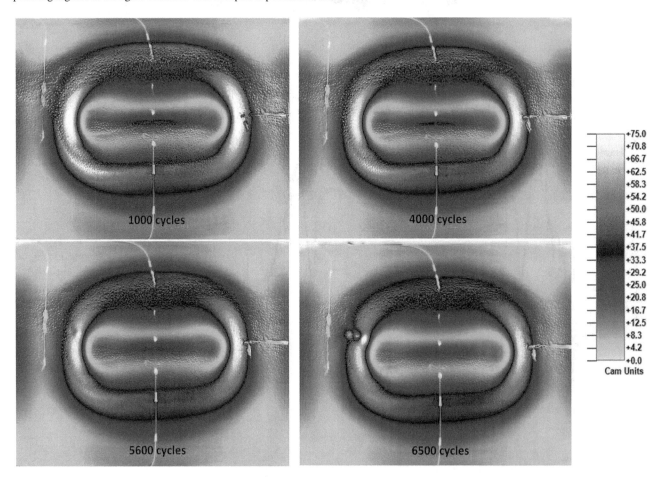

**Fig. 8.12** TSA magnitude image of the dented area taken at specific number of cycles showing crack development (after 5600 cycles) and through crack (after 6500 cycles)

$$\varepsilon_a = \left(3.5\frac{S_u - \sigma_m}{E}(N_c)^{-0.12} + \varepsilon_f^{0.6}(N_c)^{-0.6}\right)\frac{10^6}{2} \qquad (8.1)$$

The number of cycles imposed to the pipe specimen is an outcome of the internal pressure load range and respective associated circumferential strain amplitudes. These strains were used in the fatigue calculations instead of the von Mises

**Table 8.2** Fatigue life calculated circumferential based on circumferential strain amplitudes of the last cycle of the first load sequence

| Position | Fatigue life, Nc (cycles) | |
|---|---|---|
| | DIC | FE |
| A | 4613 | 4952 |
| B | 5260 | 4965 |

strains, the reason being the comparatively low value of the longitudinal strains. Using the strains presented in Fig. 8.7 for the last cycle of the first load sequence, the fatigue life was calculated and presented in Table 8.2. The calculated fatigue life results were very similar to the number of cycles that the specimen actually resisted before leaking (6584 cycles). Considering all the uncertainties associate with fatigue calculations and with the strain measurements these results can be looked as satisfactory once they are relatively close.

## Conclusions

The present paper combines a set of different experimental and numerical techniques to assess the fatigue behavior of a real scale pipe specimen containing a plain longitudinal dent. The results obtained show good agreement between them. The DIC technique not only mapped accurately the specimen shape, but also determined the overall strain field on the surface and helped determining the critical hot spots positions. The generation of a numerical indented pipe FE model using the deformed shape provided by DIC technique showed that by imputing the real defect geometry the strain and stress states at the hot-spots could be accurately evaluated without the need to actually perform actual cyclic tests. FBSG were used to monitor the real time execution of the test and its results were very alike to the ones provided by the other techniques. Another objective was achieved by showing that using a relatively low cost infrared camera and dedicated TSA software it is viable to rapid asses the fatigue properties of the specimen material by utilizing IR thermographic methods and to determine hot spots, their severity in terms of stress ranges, detect crack initiation, assess and monitor crack propagation applying TSA technique. The present findings can be applied to other structures that may present dents such as tanks and pressure vessels.

## References

1. Rosenfeld, M. (2001). *Proposed new guidelines for ASME B31.8 on assessment of dents and mechanical damage*. Topical report GRI-01/0084. Des Plaines, IL: GRI.
2. Fowler, J. R. (1993). *Criteria for dent acceptability in offshore pipeline. 25th Offshore Technology Conference*, Houston, Texas, USA, 3-6 May, 1993.
3. Garbatov, Y., & Guedes Soares, C. (2017). Fatigue reliability of dented pipeline based on limited experimental data. *International Journal of Pressure Vessels and Piping, 155*, 15–26.
4. Pinheiro, B. C., & Pasqualino, I. P. (2009). Fatigue analysis of damaged steel pipelines under cyclic internal pressure. *International Journal of Fatigue, 31*, 962–973.
5. Cosham, A., & Hopkins, P. (2004). The effect of dents in pipelines—Guidance in the pipeline defect assessment manual. *International Journal of Pressure Vessels an Piping, 81*, 127–139.
6. Fitness-for-Service, API 579-1/ASME FFS-1, June, 2016. American Petroleum Institute, American Society of Mechanical Engineers.
7. Paiva, V. E. L., Gonzáles, G. L. G., Vieira, R. D., Maneschy, J. E., Vieira, R. B., & Freire, J. L. F. (2018). Fatigue monitoring of a dented piping specimen using infrared thermography, PVP2018–84597. In *Proceedings of the ASME 2018 Pressure Vessels and Piping Conference*, July 15–20, 2018. Prague: The American Society of Mechanical Engineers.
8. Paiva, V. E. L., Gonzáles, G. L. G., Vieira, R. D., Maneschy, J. E., Freire, J. L. F., Ribeiro, A. S., & Almeida, A. L. F. S. (2019). Fatigue assessment and monitoring of a dented pipeline, PVP2019–93663. In *Proceedings of the ASME 2019 Pressure Vessels and Piping Conference*, July 14–19, 2019. San Antonio, TX: The American Society of Mechanical Engineers.
9. La Rosa, G., & Risitano, A. (2000). Thermographic methodology for rapid determination of the fatigue limit of materials and mechanical components. *International Journal of Fatigue, 22*, 65–73.
10. Fargione, G., Geraci, A., La Rosa, G., & Risitano, A. (2000). Rapid determination of the fatigue curve by the thermographic method. *International Journal of Fatigue, 24*, 11–19.
11. Risitano, & Risitano, G. (2010). Cumulative damage evaluation of steel fracture mechanics. *Theoretical and Applied Fracture Mechanics, 54*, 82–90.
12. Risitano, G., Risitano, A., & Clienti, C. (2011). Determination of the fatigue limit by semi static tests. *Convegno Nationale, IGF XXI, Cassino*, Italia 13–15 Giugo 2011 (pp. 322–330).
13. Wong, K., Sparrow, J. G., & Dunn, S. A. (1987). On the revised theory of the thermoelastic effect. In *SPIE Vol. 731 Stress Analysis by Thermoelastic Techniques 1987*.

14. Wong, K., Jones, R., & Sparrow, J. G. (1987). Thermoelastic constant or thermoelastic parameter? *Journal of Physics and Chemistry of Solids, 48*(8), 749–753.
15. Pitarresi, G., & Patterson, E. A. (2003). A review of the general theory of thermoelastic stress analysis. *The Journal of Strain Analysis for Engineering Design, 38*(5), 405–417.
16. Greene, R. J., Patterson, E. A., & Rowlands, R. E. (2008). Thermoelastic stress analysis. In W. Sharpe (Ed.), *Springer handbook of experimental solid mechanics*. Boston: Springer.
17. Dulieu-Barton, J. M., & Stanley, P. (1998). Development and applications of thermoelastic stress analysis. *The Journal of Strain Analysis for Engineering Design, 33*(2), 93–104.
18. Freire, J. L. F., Waugh, R. C., Fruehmann, R., & Dulieu-Barton, J. M. (2015). Using thermoelastic stress analysis to detect damaged and hot spot areas in structural components. *Journal of Mechanics Engineering and Automation, 5*(11), 623–634.
19. Vieira, R. B., Kurunthottikkal Philip, S., Gonzáles, G. L. G., Freire, J. L. F., Yang, B., & Rowlands, R. E. (2016). Determination of a U-notch stress concentration factor using thermoelasticity. *Journal of Mechanics Engineering and Automation, 6*(2), 66–76.
20. Sutton, M. A., Orteu, J. J., & Schreier, H. W. (2009). *Image correlation for shape, motion and deformation measurements*. New York: Springer Science + Business Media, LLC.
21. Shukla, A., & Dally, J. W. (2010). *Experimental Solid Mechanics*. Knoxville: College House Enterprises, LLC.
22. Castro, J. T. P., & Meggiolaro, M. A. (2013). *Fatigue design techniques: Vol. 1: High-cycle fatigue*. Create space. 2016. Specification for line pipe, API specification 5L, 2013. American Petroleum Institute.
23. Paiva, V. E. L., Vieira, R. D., & Freire, J. L. F. (2018), Fatigue properties assessment of API 5L Gr. B pipeline steel using infrared thermography. In *Proceedings of the 2018 Experimental and Applied Mechanics Meeting, SEM, Society for Experimental Mechanics*.
24. Specification for line pipe, API Specification 5L, 2013. American Petroleum Institute.
25. Vieira, R. B., Gonzáles, G. L. G., & Freire, J. L. F. (2018). Thermography applied to the study of fatigue crack propagation in polycarbonate. *Experimental Mechanics, 58*, 269.
26. Paiva, V. E. L., Freire, J. L., & Etchebehere, R. C. (2018). Assessment of chain links using infrared thermography. In *CONAEND & IEV2018–378, Congresso Nacional de Ensaios Não Destrutivos e Inspeção, 21ª IEV Conferencia Internacional sobre Evaluación de Integridad y Extensión de Vida de Equipos Industriales ABENDI*, 2018, São Paulo. Anais of CONAEND & IEV2018 (pp. 1–17).

# Chapter 9
# Development of Optical Technique For Measuring Kinematic Fields in Presence of Cracks, FIB-SEM-DIC

Y. Mammadi, A. Joseph, A. Joulain, J. Bonneville, C. Tromas, S. Hedan, and V. Valle

**Abstract** Currently, kinematic field measurements for studying the mechanical behavior of materials and structures use common optical methods, such as mark tracking techniques, grid methods and correlation techniques (Sutton et al., 2009, Image correlation for shape, motion and deformation measurements, Springer, Berlin, https://doi.org/10.1007/978-0-387-78747-3). These techniques are used over a region of interest ranging from micro to millimeter scale. However, when studies need to be conducted on even smaller scales such as sub-micrometric scale, the use of more complex means of observation is required. In this case the work can be achieved using the scanning electron microscope SEM, or some specific marking techniques as the Dual-Beam FIB [ANR-11-LABX-0017-01].

For this application, the Digital Image Correlation DIC is chosen to investigate the material behavior. In the present approach, an artificial speckle having the depth of the engraving around (a few hundred nanometers) 200 nm was used. Statistical evaluations such as grayscale histograms, autocorrelations, defocusing effects, and rigid body displacement effects are used to evaluate the error measurement in a field of 100 μm width. Various tests were also performed to ensure the repeatability and reproducibility of the method. The order of the errors is much greater than those obtained in classical optical conditions, but is less than 0.05 pixel.

An application to study the mechanical behavior of a metallic composite is proposed. These composites Al/ωAl-Cu-Fe (Joseph 2016) have a local behavior depending on the local material structure, which can be brittle or ductile. An adaptation of DIC method (H-DIC) is proposed to study the mechanisms of deformation at these scales, taking into account local fractures (Valle et al. 2015) or local plastic strains. The particularity of this method extension lies in the good separation of the strain fields and the cracked part. Results of a single test are presented and discussed here, which focus on the comparison between a classical DIC analysis and its extension.

**Keywords** Correlation · FIB-SEM SEM-FEG · Kinematic field · Scanning electron microscopy Digital image correlation Speckle · DIC-HDIC

## Introduction

The complete quantitative characterization of the material behavior at small-scales has recently been done to investigate nanoscale spatial resolution by combining scanning electron microscopy (SEM) and digital image correlation (DIC) [1–3]. This association, called (SEM-DIC) is an emerging and powerful technique used to study microscopic phenomena in a wide range of materials [4–6].

The first difficulty is linked to the SEM and the problem of image stability and distortion [7]. In contrast to the proven performance of the DIC methods using optical images, SEM-DIC suffers of lakes directly associated to instability and distortion of this tools.

The second difficulty, which is the same as in optical condition, is to correctly observe and separate strain localization and fracture developments. Using conventional DIC in optical condition, some works have been conducted on materials with developed cracks, but it is always difficult to localize cracks and to measure accurately strain fields in a same region. As the conventional DIC was developed on the basis of continuum mechanics, this method is clearly not adapted to the observation of cracked materials. Recently, a global approach gave interesting results [8], but this method supposes knowing

---

Y. Mammadi (✉) · A. Joseph · A. Joulain · J. Bonneville · C. Tromas · V. Valle
Institut PPRIME, University of Poitiers, CNRS, ENSMA, UPR 3346, Chasseneuil Cedex, France
e-mail: younes.mammadi@univ-poitiers.fr

S. Hedan
Institut IC2MP, University of Poitiers, CNRS, UMR 7285, ENSI Poitiers, Poitiers Cedex, France

the crack position before the image analysis. A more recent work proposed a local approach, adding a specific enrichment of the kinematical field, taking into account the presence of cracks (Heaviside-DIC) [9, 10]. This method has the advantage of analysing images without knowing the crack positions. As the proposed enrichment of this method is added to a classical kinematical field using first gradients of displacement, it can be possible to separate strains from cracks.

The aim of the proposed paper is to:

- Define the process of marking and imaging using Focused Ion Beam (FIB) and SEM equipment, and then analyze displacement measurement performances.
- Describe an optimized process for strain and discontinuous displacement measurements.
- Investigate these developments for the study of a metallic composite material.

## Experimental Procedures

Since the position of the electron beam is not a controlled parameter in typical modern SEM systems and because of various environmental factors, positioning errors will occur during the scanning process (imaging). Some researchers have focused on the effect of the resulting time-dependent drift variations on the secondary electronic signal (SE) which is most often used for metrological studies [2]. In practice, it is necessary to evaluate the performances of the imaging process, using patterns linked to the used metrology. In the present work, the Digital Image Correraltion was used and the adapted pattern was chosen as a speckle. Two aspects are involved in the design of the speckle pattern. The speckle morphology and the grain size directly influences the quality of the results and determines the sensitivity to movement. Many techniques exist to add a speckle on a specimen, like chemical attack, particle deposition, or engraving. One of these method offering the possibility to control the speckle characteristics is to use a FIB system to engrave the surface of the specimen., Using this system, it is easy to control the morphology and the magnification (the depth) of the generated speckle, directly from a synthetic image file. Therefore, the choice of the optimal model and the good magnification of the FIB, condition the design of the speckle pattern and the response to the measurement requirements [11].

### Making Speckles by FIB

In the presented work, a FIB was used to produce a controlled speckle. In our experiments, FIB Helios field FIB-SEM Fig. 9.1 is employed with milling parameter acceleration voltage equal to 30 KeV and beam current equal to 2.5 nA. The synthetic

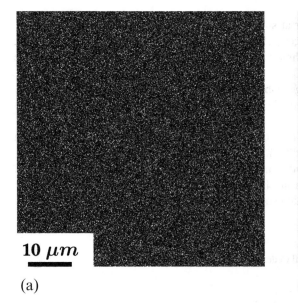
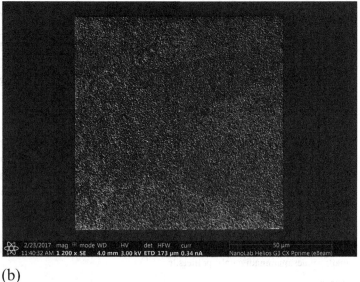

(a) (b)

**Fig. 9.1** (a) Speckle pattern and (b) example using FIB etching Inconel 718

speckle pattern used for the milling of the specimen is shown Fig. 9.1a and the corresponding milled surface of the specimen is presented on Fig. 9.1b. Using FIB milling to create speckle patterns not only extends the application of FIB but also offers a new patterning technique for DIC method. An Inconel 718 test specimen with a smooth surface was used as a substrate and the resulting depth of the milling was evaluated by Atomic Force Microscope (AFM) and gives an average value of 200 nm.

## *Good Practice of SEM Imaging for Metrology*

When handling SEM, it is necessary to ensure a good contrast in the image of the created pattern while avoiding its saturation. There are two typical modes for SEM imaging: secondary electron imaging (SEI) and electronic backscatter imaging (BEI), In our case, leading the literature [6], the secondary electron imaging has been chosen.

A confrontation of the imaging process used from the FIB-SEM microscope used for milling, and a SEM-FEG microscope was proposed. The diffraction patterns were acquired with the FIB-SEM microscope using an acceleration voltage of 3 KeV, a beam current of 0.34 nA and Working Distance (WD = 4 mm) Fig. 9.1. For the SEM-FEG microscope a voltage of 15 KeV, a beam curent of 1.5 nA and a WD = 10 mm were used. The principal difference between a SEM imaging and a SEM quantitative imaging for metrology, is to respect the same principle applied using optical system. The main important condition to do a quantitative imaging is to avoid modifications of the microscope parameters. For example, if there a modification of the focus to do, it is better to use the microscope displacement stage (Z direction) instead of using the electronic adjustment.

## *Displacement Error Evaluations*

Various tests have been made to measure the repeatability and reproducibility of the method. Horizontal and vertical displacements have been imposed and the error related to these displacements were computed Figs. 9.2 and 9.3.

Figure 9.2 shows the results issued from the DIC displacement analysis for the FIB-SEM microscope imaging. Figure 9.3 shows the same analysis from the SEM-FEG microscope imaging.

On these two experiments, the same results can be observed. The displacement field is not uniform along the Y axis, which is the well-known drift phenomenon of the SEM systems. The amplitude of this phenomenon can be evaluated less than 0.4 pixels. However, the local error (standard deviation) shows a lower level (<0.05 pixel). Both these results show that the evaluation of the displacement by SEM remains a difficulty because of global errors, but is less important if one focuses on the strains. Let us recall that in optical metrology the local error using DIC, is near the global one, and is about 0.02 pixels.

The comparison of the two tools does not show any notable differences. For the rest of the study, the SEM-FEG microscope was used for the image acquisitions. It is interesting to evaluate the stability of the SEM systems. This stability is very essential

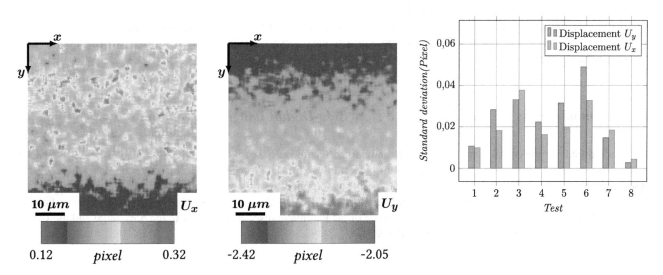

**Fig. 9.2** Cartography of the displacements $u_x$ & $u_y$ and corresponding measurement errors from analysis (FIB-SEM)

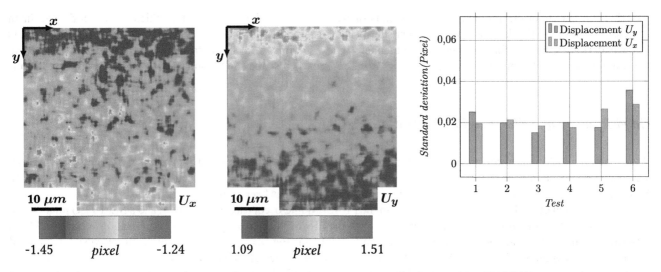

**Fig. 9.3** Cartography of the displacements $u_x$ & $u_y$ and corresponding measurement errors from analysis (SEM-FEG)

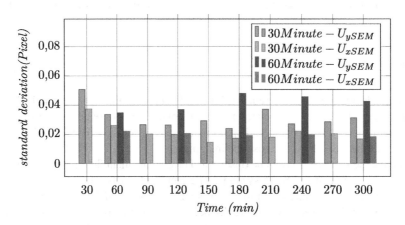

**Fig. 9.4** Stability errors as a function of time for 30–60 min spaced points

if the experiments need a long time process. Two evaluations of the displacement errors have been done with different steps (30 min and 1 h), and with a total duration of 4 h.

On Fig. 9.4, it can be seen that measurement errors do not change over time and stay less than 0.05 pixels even after 4 h Fig. 9.4. This is a very interesting result because, usally in-situ measures are considered for long term experimentation while this shows that even ex-situ measures can be taken into consideration.

## *Separation of Cracks and Strains Using a Heaviside-Digital Image Correlation Method*

Heaviside-digital image correlation method [9] is an adapted technique based on local DIC approach to be able to extract cracks using an enrichment of the kinematical transformation used in DIC process 9.1. This method was used to capture cracks without knowing their position and orientation and was applied to different materials and in different domains [10]. Where $\underline{X}$ the position of the subset in the reference image, $(\underline{X} - \underline{X_0})$ is the position in the subset and $\underline{U}$ is a translation in the plane. The jump/step function defines the magnitude of the kinematic discontinuity in both the horizontal and the vertical directions represented by $\underline{U}'$ reference frame in Fig. 9.5.

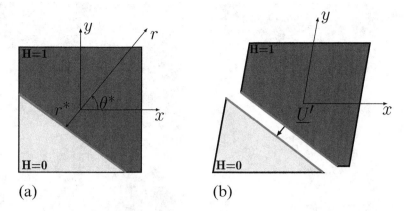

**Fig. 9.5** Subset representation of the effect of Heaviside function from the initial (**a**) to the final (**b**) state. The kinematical transformation is composed by a rigid boby movement (not represented here) a displacement jump $\underline{U'}$ and a deformation

The kinematical transformation can be expressed using an enrichment based on a 2D Heaviside function $H$.

$$X^* = \underbrace{\underline{X} + \overbrace{\underline{U}}^{\text{Rigid body}} + \overbrace{\frac{\partial U}{\partial X}(\underline{X} - \underline{X_0})}^{\text{First gradient}}}_{\text{Conventional DIC method}} + \underbrace{\overbrace{\underline{U'}}^{\text{Jump}} \times \overbrace{H(\underline{X} - \underline{X_0})}^{\text{Heaviside function}}}_{\text{Heaviside - DIC method}} \quad (9.1)$$

where $\underline{X}$ are the coordinates of a point in the initial image, $\underline{U}$ the rigid body displacement vector, $\frac{\partial U}{\partial X}$ the gradient tensor giving the strain tensor, $\underline{X_0}$, the center of the subset and $\underline{U'}$ the discontinuity jump vector.

A pseudo strain indicator $\varepsilon_{eq}$ was calculated in the spirit of Tresca to show the strain amplitude:

$$\varepsilon_{eq} = \frac{|\varepsilon_1 - \varepsilon_2|}{2} \quad (9.2)$$

We shall now present an application on a material undergoing important deformations. This study was carried out with an external loading system (ex-situ study). In this study, the position of the sample should be retrieved in the SEM at each loading step. Following The previous recommendations, it was necessary during this study to not modify the settings of the SEM microscope between the initial image and the images gotten from each loading step.

## Application and Analysis on a Composite Metallic Material Al/ωAl-Cu-Fe

### Material Global Characteristics

This material is a two phases composite which have two different behaviors which makes the study more difficult. The Al phase is less hard than the $\omega$-Al-Cu-Fe phase Fig. 9.6. Indeed, by observing the speckle profile on the AFM, it can be noticed that the roughness is less important on the phase omega-Al-Cu-Fe than on the phase Al.

### Marking Process and Specimen Dimension

Several attempts have been made to find the correct depth of milling of the Al/$\omega$-Al-Cu-Fe biphase. The dimension of the specimen is $6.25 \times 2.5 \times 2.5$ mm$^3$ and it is loaded in compression. Subsequently, about ten $100 \times 100$ μm zones were marked with a speckle but only one of these zones of analysis was presented.

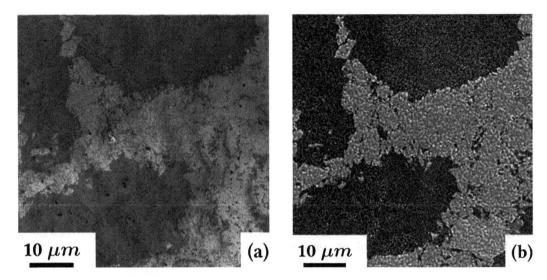

**Fig. 9.6** (a) Zone before FIB patterning, and (b) Zone after FIB milling

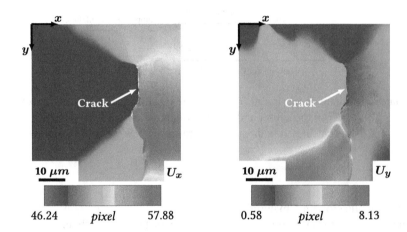

**Fig. 9.7** Measured displacement components from Heaviside-DIC computation for one zone of the second unloading step

## First Results

It can be observed on Fig. 9.6a that the specimen surface gives different aspects linked to the different phases of the material. On Fig. 9.6b, it can be shown that the phase hardness influences the machining depth, and thus, the marking contrast. Over the phase $\omega$, the maximum and the minimal values of the depth are respectively 400 and 110 nm. The Al phase has a lower amplitude with a maximal and a minimal depth values of 300 and 220 nm Fig. 9.6.

After each step of loading, the specimen is replaced at the same position in the SEM, and each zone is retrieved. As it is difficult to retrieve the exact positions and orientations, the presented displacement maps are not representatives of the real movement of each zone in the specimen base. Only jump maps and strain maps give relevant data. However, a single pair of measured displacement components is plotted in Fig. 9.7 for one zone, showing the ability to capture unperturbed maps, with discontinuities and visible gradients.

Multiple loading steps were conducted on the same sample, and the one chosen to be presented here was considered the most interesting. During previous loading states, micro cracks appeared on different zones of the sample, those cracks helped majorly the development of the actual main crack appearing in the image.

In this step of loading, the strains and the cracks are well developed. It can be observed that the cracks are localized in the $\omega$ phase, while the strains are concentrated in the Al phase. This observation confirms that the $\omega$ phase has a brittle behavior while the Al phase has a ductile behavior. It can be noted that the local plasticity is developed from cracks, and can link, for example, two cracks. It can also be observed the presence of a crack along the interface between $\omega$ phase and Al phase can also be observed Fig. 9.8.

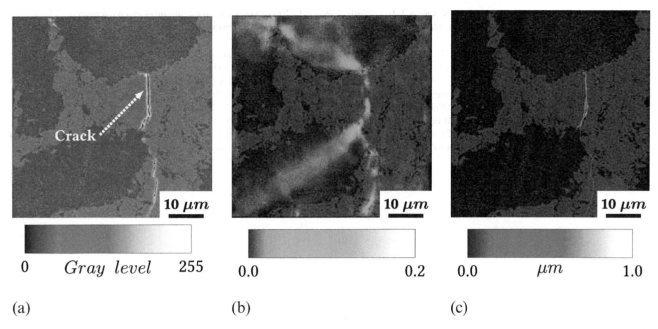

**Fig. 9.8** Maps for the second unloading step. (**a**) SEM image, (**b**) equivalent strain $\varepsilon_{\text{eqHDIC}}$ and (**c**) magnitude of displacement jump vector $||\underline{U'}||$

## Conclusions

This work presents, a development of an optical technique for measuring kinematic fields in presence of cracks, FIB-SEM-DIC. First results presented displacement error evaluations. The accuracy of the displacement linked to the stability of the SEM was evaluated to 0.05 pixels. After a brief study on the investigated material, the HDIC method were employed to solve the problem of kinematical discontinuities that result from plastic deformation and challenge conventional digital image correlation method (DIC).

After this first validation, the specimen was etched. Two different behaviors of the two different phases, were observed highlighting the brittle behavior of the omega phase and the ductile behavior of the Al phase. This test demonstrates that the HDIC method coupled with FIB etching and SEM Imaging is able to detect fractures, and to separate then from strains. It gives the opportunity to understand better how strains and cracks are linked to each other, in a complex material under mechanical straining.

**Acknowledgements** The authors gratefully acknowledge the support of LABEX Global. This work partially pertains to the French Government program "Investissements d'Avenir". (LABEX INTERACTIFS, reference ANR-11-LABX-0017-01). This work has been partially supported by Nouvelle Aquitaine Region and by European Structural and Investment Funds (ERDF reference: P-2016-BAFE-94/95).

## References

1. Sutton, M. A., Wolters, W. J., Peters, W. H., Ranson, W. F., & McNeill, S. R. (1983). Determination of displacements using an improved digital correlation method. *Image and Vision Computing, 1*(3), 133–139. https://doi.org/10.1016/0262-8856(83)90064-1.
2. Sutton, M. A., Li, N., Joy, D. C., Reynolds, A. P., & Li, X. (2007). Scanning electron microscopy for quantitative small and large deformation measurements Part I: SEM imaging at magnifications from 200 to 10,000. *Experimental Mechanics, 47*(6), 775–787. https://doi.org/10.1007/s11340-007-9042-z.
3. Sutton, M. A., Schreier, H., & Orteu, J. J. (2009). *Image correlation for shape, motion and deformation measurements*. Berlin: Springer. https://doi.org/10.1007/978-0-387-78747-3.
4. Kammers, A. D., & Daly, S. (2011). Small-scale patterning methods for digital image correlation under scanning electron microscopy. *Measurement Science and Technology, 22*(12), 125501. https://doi.org/10.1088/0957-0233/22/12/125501.
5. Kammers, A. D., & Daly, S. (2013). Digital Image correlation under scanning electron microscopy: Methodology and validation. *Experimental Mechanics, 53*(9), 1743–1761. https://doi.org/10.1007/s11340-013-9782-x.
6. Zhu, R., Xie, H., Xue, Y., Wang, L., & Li, Y. (2015). Fabrication of speckle patterns by focused ion beam deposition and its application to micro-scale residual stress measurement. *Measurement Science and Technology, 26*(9), 095601. https://doi.org/10.1088/0957-0233/26/9/095601.

7. Sutton, M. A., Li, N., Garcia, D., Cornille, N., Orteu, J. J., McNeill, S. R., et al. (2006). Metrology in a scanning electron microscope: Theoretical developments and experimental validation. *Measurement Science and Technology, 17*(10), 2613–2622. https://doi.org/10.1088/0957-0233/17/10/012.
8. Bertin, M., Du, C., Hoefnagels, J. P., & Hild, F. (2016). Crystal plasticity parameter identification with 3D measurements and integrated digital image correlation. *Acta Materialia, 116*, 321–331. https://doi.org/10.1016/j.actamat.2016.06.039.
9. Valle, V., Hedan, S., Cosenza, P., Fauchille, A. L., & Berdjane, M. (2014). Digital image correlation development for the study of materials including multiple crossing cracks. *Experimental Mechanics, 55*(2), 379–391. https://doi.org/10.1007/s11340-014-9948-1.
10. Bourdin, F., Stinville, J. C., Echlin, M. P., Callahan, P. G., Lenthe, W. C., Torbet, C. J., et al. (2018). Measurements of plastic localization by heaviside-digital image correlation. *Acta Materialia, 157*, 307–325. https://doi.org/10.1016/j.actamat.2018.07.013.
11. Li, Y., Xie, H., Luo, Q., Gu, C., Hu, Z., Chen, P., et al. (2012). Fabrication technique of micro/nano-scale speckle patterns with focused ion beam. *Science China Physics, Mechanics and Astronomy, 55*(6), 1037–1044. https://doi.org/10.1007/s11433-012-4751-4.

# Chapter 10
# DIC Determination of SIF in Orthotropic Composite

**N. S. Fatima and R. E. Rowlands**

**Abstract** Based on original analytical concepts by Khalil et al. (International Journal of Fracture 31:37–51, 1986) and subsequently augmented by Ju and Rowlands (Journal of Composite Materials 37:2011–2025, 2003) to study inclined cracks thermoelastically in orthotropic plates, this paper determines the stress intensity factor (SIF) in a double-edge cracked finite orthotropic tensile composite (Fig. 10.1) from a single digital image correlation (DIC) recorded displacement field. The traction-free condition of the crack-face was accounted for using conformal mapping.

**Keywords** Stress intensity factor · Composite · DIC · Double edged matched plate

## Introduction

The present technique offers a simple and effective means of analyzing orthotropic structures containing cracks and evaluating fracture related important parameters such as the stress intensity factor (SIF). The method is applicable to anisotropic plates with various types of linear cracks under mode-I and mode-II loading. The analytical concepts were originally proposed by Khalil et al. [1] in their hybrid finite element method to analyze anisotropic materials with cracks. Later the technique was numerically and experimentally utilized by Ju [2] and Ju and Rowlands [3], respectively, and extended for plates with inclined cracks. In this present study Khilil's analytical approach, which employs the Airy stress functions with conformal mapping, is combined with numerical tools to process a single component of displacement measured by digital image correlation to evaluate the SIF in a double-edge crack, finite ($2a/W = 0.6$), $[0_{13}/90_5/0_{13}]$ graphite-epoxy plate ($E_{11} = 104.1$ GPa, $E_{22} = 28.1$ GPa, $\nu_{12} = 0.155$ and $G_{12} = 3.0$ GPa), with the 1-direction in the vertical $y$-direction, Fig. 10.1. Measured displacement information was considered away from the crack, which is a significant advantage for practical applications as measured information near a discontinuity is often unreliable. The method also requires no knowledge of the external loading.

## Analytical Background

For the coordinate system with the sharp crack along the negative portion of the $z$-plane (Fig. 10.1a) and the crack-tip at $x = y = z = 0$, the displacement field near the crack-tip in an anisotropic material can be expressed as

$$v = \sum_{j=1}^{2N} E_j(z_1, z_2)\, \gamma_j \qquad (10.1)$$

where $z_j = x + \mu_j y$ for $j = 1, 2$. The stress coefficients, $\gamma_j$, are evaluated by DIC. Knowing $\gamma_j$, $K_\mathrm{I}$ is available from

$$K_\mathrm{I} = \sqrt{2\pi}\left[\left(1 - \frac{\beta_1}{\beta_2}\right)\gamma_1 + \left(\frac{\alpha_2 - \alpha_1}{\beta_2}\right)\gamma_{1+N}\right] \qquad (10.2)$$

Values of $E_j$, $\alpha_j$ and $\beta_j$ depend on the composite's constitutive properties.

---

N. S. Fatima · R. E. Rowlands (✉)
Department of Mechanical Engineering, University of Wisconsin-Madison, Madison, WI, USA
e-mail: rowlands@engr.wisc.edu

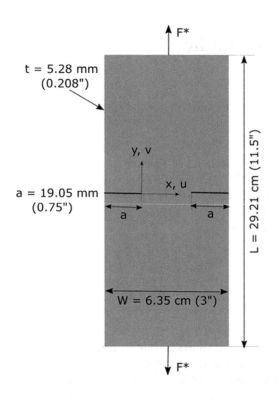 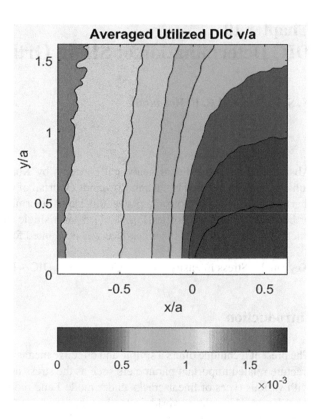

(a) (b)

**Fig. 10.1** (a) Double-edge crack $[0_{13}/90_5/0_{13}]$ graphite-epoxy plate and (b) DIC-recorded vertical displacements in neighborhood of left crack-tip

The technique employs the mathematical concepts of complex variables Airy stress functions combined with conformal mapping which satisfies the crack-tip stress singularity. Equation (10.3) maps the half-plane from the $\zeta$-plane into an edge crack in the physical $z$-plane [1, 4]

$$z_j = \omega_j\left(\zeta_j\right) = \zeta_j^2, \quad j = 1, 2 \tag{10.3}$$

The complex material properties, $\mu_j = \alpha_j \pm i\beta_j$, used in the expression of the complex variables, $z_j = x + \mu_j y$ for $j = 1$, 2, are obtained from following equation [5].

$$a_{11}\mu^4 - 2a_{16}\mu^3 + (2a_{12} + a_{66})\mu^2 - 2a_{26}\mu + a_{22} = 0 \tag{10.4}$$

where $a_{ij}$ are the elastic compliances of the orthotropic material and for a plate with the crack-face normal to the material's strong/stiff direction and loaded along one of the principal material directions are defined as

$$a_{11} = \frac{1}{E_{11}}, a_{12} = \frac{-v_{21}}{E_{22}} = \frac{-v_{12}}{E_{11}}, a_{22} = \frac{1}{E_{22}}, a_{66} = \frac{1}{G_{12}}, a_{16} = a_{26} = 0 \tag{10.5}$$

The near crack-tip displacement, $U_k$, and stress, $V_k$, fields in an anisotropic material can be expressed as [1–3]

$$U_k = \sum_{j=1}^{2N} E_j(z_1, z_2)\gamma_j \text{ for } k = 1, 2 \tag{10.6}$$

$$V_k = \sum_{j=1}^{2N} A_j(z_1, z_2)\gamma_j \text{ for } k = 1, 2, 3 \tag{10.7}$$

The $\gamma_j$ are stress coefficients and $N$ is the number of displacement terms retained, where for $N = 1, j = 2$ $N = 2$, i.e., for one displacement term there are two stress coefficients, $\gamma_j$ and $\gamma_{j+N}$. As the following equations show, terms $E_j$ and $A_j$ are functions of the orthotropic material properties and coordinate locations [2]

$$E_j = 2\text{Re}\left(s_1 z_1^{\frac{j}{2}} + s_2 M_{1j} z_2^{\frac{j}{2}}\right) \tag{10.8}$$

$$E_{j+N} = 2\text{Re}\left(i s_1 z_1^{\frac{j}{2}} + s_2 M_{2j} z_2^{\frac{j}{2}}\right) \tag{10.9}$$

For $k = 1$ in Eq. (10.6) $U_1 = u$, whereas in Eqs. (10.8) and (10.9) $s_1$ and $s_2$ are $p_1$ and $p_2$, respectively, and for $k = 2$, $U_2 = v$, $s_1$ and $s_2$ are $q_1$ and $q_2$, respectively. The complex material properties $p_1, p_2, q_1$ and $q_2$ are defined as [5]

$$\begin{aligned} p_1 &= a_{11}\mu_1^2 + a_{12} - a_{16}\mu_1 \quad p_2 = a_{11}\mu_2^2 + a_{12} - a_{16}\mu_2 \\ q_1 &= a_{12}\mu_1 + \frac{a_{22}}{\mu_1} - a_{26} \quad q_2 = a_{12}\mu_2 + \frac{a_{22}}{\mu_2} - a_{26} \end{aligned} \tag{10.10}$$

Quantities $M_{1j}, M_{2j}, A_j$ and $A_{j+N}$ are given by [1–3]

$$\begin{aligned} M_{1j} &= -\frac{\beta_1}{\beta_2} \text{ and } M_{2j} = \frac{\alpha_2 - \alpha_1}{\beta_2} - i \text{ for odd values of } j \\ M_{1j} &= -1 + i\frac{\alpha_1 - \alpha_2}{\beta_2} \text{ and } M_{2j} = -i\frac{-\beta_1}{\beta_2} \text{ for even values of } j \end{aligned} \tag{10.11}$$

$$A_j = (-1)^{k-1} j \, Re\left(\mu_1^{k-1} z_1^{\frac{j-2}{2}} + \mu_2^{k-1} M_{1j} z_2^{\frac{j-2}{2}}\right) \tag{10.12}$$

$$A_{j+N} = (-1)^{k-1} j \, Re\left(i\mu_1^{k-1} z_1^{\frac{j-2}{2}} + \mu_2^{k-1} M_{2j} z_2^{\frac{j-2}{2}}\right) \tag{10.13}$$

For the stress components, $k = 1, 2, 3$ in Eqs. (10.7), (10.12) and (10.13) represents $\sigma_{yy}$, $\sigma_{xy}$ and $\sigma_{xx}$, respectively. The SIF, $K_I$, in Eq. (10.2) at locations very close to the crack-tip, is dominated by the two stress coefficients $\gamma_1$ and $\gamma_{1+N}$, which can be readily obtained using measured displacement data in Eq. (10.1).

## Analytical-Experimental Hybridization

For $k = 2$ in Eq. (10.6), Eq. (10.1) is obtained. Further re-arranging Eq. (10.1) and writing in a matrix form provides Eq. (10.14), where values of $E_j$'s are known from Eqs. (10.8) through (10.11). Measured values of displacements in the loading direction, $v$, (Fig. 10.1b) are used to reduce the number of unknowns in Eq. (10.14). Equation (10.14) can again be written as Eq. (10.15) where for $n$ measured values of displacements, $\{d\}$, the only unknown are the $2N$ number of stress coefficients, $\{g\}$. The over-determined system of linear Eqs. (10.15) can be solved for the unknown stress coefficients, $\gamma_j$ and $\gamma_{j+N}$, using least squares [6]. The plate of Fig. 10.1a was loaded in a hydraulic-grip MTS loading-frame and Correlated Solution, Inc.'s commercial DIC package was used to obtain the displacement information. Displacements in Fig. 10.1b are those in the loading direction, normal to the crack-face, and for a vertical uniaxial tensile load of 7.1 kN (1600 lbs) [6].

In terms of the far-field stress, $\sigma_0$, and edge crack length, $a$, the presently obtained SIF value is $\frac{K_I}{\sigma_0\sqrt{a}} = 2.08$. This result was verified independently, including a comparison with that evaluated using a different technique by Sih, Paris and Irwin [6, 7].

$$\{v\} = [E_1 \ E_2 \ \ldots \ E_N \ E_{N+1} \ \ldots \ E_{2N-1} \ E_{2N}] \begin{Bmatrix} \gamma_1 \\ \gamma_2 \\ \vdots \\ \gamma_N \\ \gamma_{N+1} \\ \vdots \\ \gamma_{2N-1} \\ \gamma_{2N} \end{Bmatrix} \quad (10.14)$$

$$\{d\}_{n \times 1} = [E]_{n \times 2N} \{g\}_{2N \times 1} \quad (10.15)$$

## Summary and Conclusions

The present technique can reliably analyze finite orthotropic plates with various types of crack configurations. Equilibrium, compatibility and near crack-tip singularity are satisfied analytically. The method involves only a single component of measured information and requires no knowledge of the external loading, far-field boundary conditions or measured information near the crack-tip. These features are significant for real-world applications. The method can be utilized for evaluating both $K_\mathrm{I}$ and $K_\mathrm{II}$.

## References

1. Khalil, S. A., Sun, C. T., & Hwang, W. C. (1986). Application of a hybrid finite element method to determine stress intensity factors in unidirectional composites. *International Journal of Fracture, 31*, 37–51.
2. Ju, S. H. (1996). Simulating stress intensity factors for anisotropic materials by the least-squares method. *International Journal of Fracture, 81*, 283–297.
3. Ju, S. H., & Rowlands, R. E. (2003). Thermoelastic determination of KI and KII in an orthotropic graphite-epoxy composite. *Journal of Composite Materials, 37*, 2011–2025. https://doi.org/10.1177/0021998303036246.
4. Tong, P. (1977). A hybrid crack element for rectilinear anisotropic material. *International Journal for Numerical Methods in Engineering, 11*, 377–403. https://doi.org/10.1002/nme.1620110211.
5. Lekhnitskii, S. G. (1968). *Anisotropic plates*. New York: Gordon and Breach.
6. Fatima, N. S. (2019). *Hybrid photomechanical analyses of isotropic and orthotropic composite structures containing various geometric discontinuities*. PhD Thesis, University of Wisconsin, Madison.
7. Sih, G. C., Paris, P. C., & Irwin, G. R. (1965). On cracks in rectilinearly anisotropic bodies. *International Journal of Fracture Mechanics, 1*, 189–203. https://doi.org/10.1007/BF00186854.

# Chapter 11
# Determining In-Plane Displacement by Combining DIC Method and Plenoptic Camera Built-In Focal-Distance Change Function

Chi-Hung Hwang, Wei-Chung Wang, Shou-Hsueh Wang, Rui-Cian Weng, Chih-Yen Chen, and Yu-Chieh Chen

**Abstract** In this study, the possible application of plenoptic camera for DIC displacement is explored. A special function, focal-distance change function, of plenoptic camera is first investigated by using 2D DIC method to know how the function would change the images in size and to know the possible pseudo in-plane displacement might introduced. Then the focal-distance change function is applied to generate two images at different focal-length with respect to an image-pair taken before and after object moved, those four images can create six independent in-plane displacement fields by DIC. In this study, four displacement fields are selected out of the six independent ones and used to calculate displacement filed with smaller standard deviation. By properly linearly combining the selected displacement fields, both standard deviation of u- and v-displacement are successfully reduced to 1/3 with respect to the original values and the u- and v-displacement field become flat as compared to the original one but the mean displacement is 10% larger than the nominal displacement determined by a precision linear stage.

**Keywords** Plenoptic camera · Light field camera · Digital image correlation · In-plane displacement · Focal length

## Introduction

Digital image correlation method is an optical strain measurement method which can be implemented for determining the displacement/deformation field of an object subjected to external forces by using two digital images captured at two statuses with nature and artificial characteristic pattern on the object. For DIC method, the digital images are not limited to be taken by optical imaging systems; in fact, based on different application, the digital images can be generated by SEM, TEM, X-ray and CT. Knowing the DIC method can be used with various digital "images"; in this study, the plenoptic camera, also known as light-filed camera, is implemented to taken the images of an object at different status to determine the displacement filed. Different from traditional CCD or CMOS camera, the plenoptic camera is consist of an optical lens array placed in front of the optical sensor which enables the plenoptic camera can provide disparity map of an image. Therefore, a plenoptic camera can work as a typical digital camera; on the other hand, while the disparity maps are applied to the images, new images can be numerically generated with different focal-distance and/or depth of field.

In this study, the images of an aluminum plate with randomly speckle sprayed were first taken before and after displaced along $x$-axis; then a set of images can be numerically generated by numerically shifting the focal-length by $\Delta f$. Then, two image sets can be used for determining the in-plane displacement; the first set is images taken before and after physical in-plane displacement, and the second image set is images with an object is firstly pseudo out-of-plane displaced and then in-plane displaced. For a single displacement, four images can be used for DIC displacement determination; in principle, six independent displacement field can be calculated; in this study, the final in-plane displacement field is determined by selecting different images from those four images and calculating the displacement field by 2D DIC.

---

C.-H. Hwang (✉) · R.-C. Weng · C.-Y. Chen · Y.-C. Chen
Taiwan Instrument Research Institute, National Applied Research Laboratories, Hsinchu, Taiwan
e-mail: chhwang@narlabs.org.tw; cian@itrc.narl.org.tw; bjamesh@tiri.narl.org.tw

W.-C. Wang · S.-H. Wang
Department of Power Mechanical Engineering, National Tsing Hua University, Hsinchu, Taiwan
e-mail: wcwang@pme.nthu.edu.tw

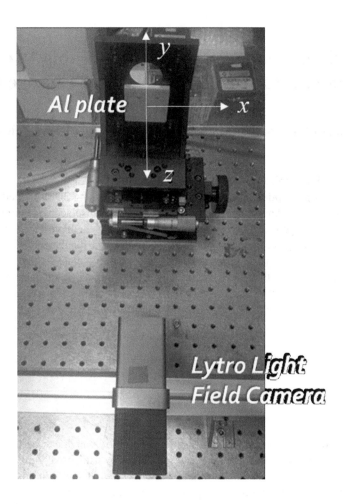

**Fig. 11.1** Experimental setup

## Experimental Setup

In this study, as shown in Fig. 11.1, an aluminum plate with artificial speckled pattern on the surface is prepared by typical spraying procedure and then mounted to a stage which can moved toward to the right. A budget plenoptic camera from Lytro Ltd., was implemented as imaging camera. Different from traditional imaging system, for a plenoptic camera, there is a micro-lens array placed in front of the detector. The budget plenoptic camera functions with an $8\times$ optical zoom lens and associated $f/2$ aperture and 1.1Mpixels CMOS detector. The micro-lens array of a plenoptic camera can be considered as a sub-aperture imaging lens, the field of view defined by the main optic is then divided into sub-images and then the image triangulation measurement is implemented to the adjacent sub-apertures to create the disparity map of an object space. Figure 11.2 is a typical images obtained by a shot of a plenoptic camera; Fig. 11.2a is an image taken by the plenoptic camera as a traditional camera can do; Fig. 11.2b is the associated disparity map of Fig. 11.2a and the gray level indicates the distance from camera. Since the plenoptic camera can provide depth information for objects in an image which enables numerical focal distance and depth of field adjustment after images taken. In this study, all the in-plane-displacement fields are then calculated by using VIC-2D from Correlated Solutions.

## Results and Discussions

Two experiments are performed in this study to explore the feasibility to implement plenoptic camera together with DIC method for displacement measurement. The first experiment is regarding to change the focal length numerically and then determine the in-plane pseudo-displacement field to know whether the focal length changes would conduct unexpected issues, such as the image is distorted/shifted or even other unexpected problems after focal length numerically changed. The second

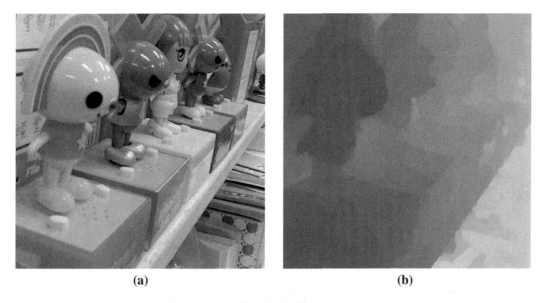

**Fig. 11.2** Information obtained by using a plenoptic camera. (**a**) Full depth of field image, (**b**) Associated disparity map

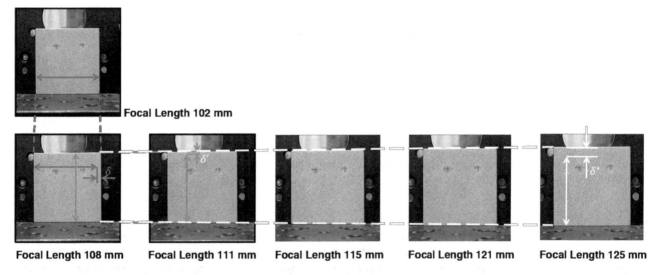

**Fig. 11.3** A series images generated numerically by changing focal length and the dimension of image are obviously changed

experimental is to explore whether it is possible to improve the determined the displacement filed by using plenoptic camera with changing focal length function.

## *Image Dimension-Change Introduced by Adjusting Focal Length Numerically*

To evaluate the dimension change caused by numerically changing the focal length of plenoptic camera, an image was first taken at 102 mm focal length and then numerically generating a series images with different focal lengths by using the software from camera provider. As shown in Fig. 11.3, a series images with focal length 108, 111, 115, 121 and 125 mm are generated from the given images; obviously, the size of the aluminum plate is enlarged as the numerical focal length is increased. In order to know the in-plane dimension changed introduced by focal length change, taken image captured at 102 mm focal length as reference, the associated in-plane dimension change for numerically generated images are calculated. The calculation is first performed by 2D DIC method with VIC-2D, the typical displacement fields are shown in Fig. 11.4. The calculated displacement filed shows that the in-plane displacement for both $u$ and $v$ fields are symmetrical with respect to $y$-axis and $x$-axis respectively which indicates the numerically generated images are expended with respect to a center. Then

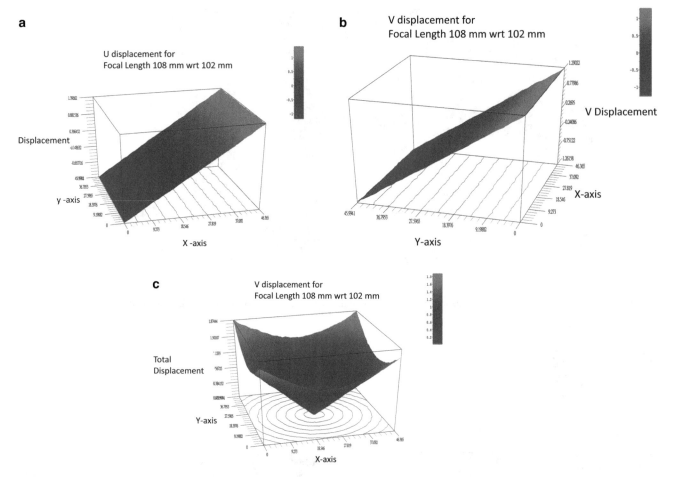

**Fig. 11.4** Pseudo in-plane displacement introduced by numerically changing the focal length. (**a**) Displacement u, (**b**) Displacement v, (**c**) Total Displacement

the dimension changes can be evaluated simply by calculating the displacement along x-axis and y-axis, then taking average for further discussion. As illustrated in Fig. 11.5, the focal length difference, $\Delta f$ ($\Delta f$ = adjusted focal length − 102 mm) and the change of in-plane dimensions are linearly correlated. Meanwhile, the geometrical centers of the aluminum plate for different images are also determined by determined displacement fields. Although the displacement fields are symmetrical with respect to y-axis and x-axis respectively, the non-displacing points along x-axis and y-axis can be evaluated by linear interpolation; the center-deviation of all numerically generated images with respect to the reference image are within [0.24pixels, 0.86pixels], and there is no obvious relation between the center-deviation and $\Delta f$. From the results, the images generated by changing focal lengths can be considered as moving the object towards a camera, as showing in Fig. 11.6. According to the imaging model, the dimension change to the moving distance should also be linear related. Unfortunately, the optical parameter of the plenoptic camera are not provided by manufacturer, additional experiment is needed to obtain the relation for dimension change to the moving distance, and this is out of the scope of this study and would not be discussed.

## Determining In-Plan Displacement by Combining Images of Different Focal Lengths

According to discussion in previous section, numerically changing the focal length is equivalent to move the object towards/away from a plenoptic camera with the image center might be slightly moved, the result reveals, by numerically changed the focal lengths, additional out-of-plane displacement fields can be added into the images captured by plenoptic camera without introducing in-plane displacement severely. Therefore, whenever a pair of images are taken before and after moved an object along z-axis, additional image-pairs can be generated numerically with an additional pseudo out-of-plane displacement by using focal length changing function of a plenoptic camera; however, in this study the accurate out-of-

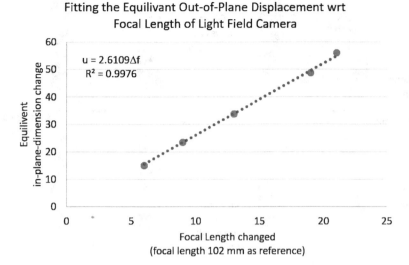

**Fig. 11.5** Relation for image dimension change and the numerical focal length change

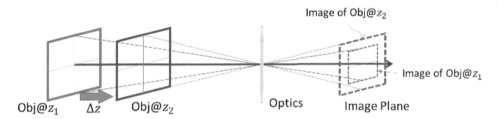

**Fig. 11.6** Simplified model for dimension change

plane displacement is unknown due to optical parameters are not provided by manufacturer; therefore, in this study, all the discussions would be based on focal length instead of numerically generated out-of-plane movement. As shown in Fig. 11.7, for an image-pair captured by a plenoptic camera before and after displacement, by numerically changing the focal length, a corresponding image-pair can be generated that means four images can be used for displacement evaluation. Considering DIC can determine displacement filed from two images captured at different displacement status, therefore, 12 displacement fields ($D_{ij}$, $i, j = 1, 2, 3, 4$) can be calculated from four different images. However, considering the displacement field for two images determined by DIC would change sign when the reference image and objective image are exchanged, i.e., $D_{ij} = -D_{ji}$; therefore, only six different displacement fields can be obtained. While the aluminum plate is moved 3 mm to the right, the corresponding components of displacements for all $D_{ij}$ are tabulated in Table 11.1; where $\Delta z$ is pseudo out-of-displacement introduced by changing focal length, as mentioned previously, $\Delta z$ can be determined with an additional measurement but would not be measured here. As indicated in previous section, changing focal length can be equilivent to out-of-plane displacement with image center slightly shifted; therefore, displacement field $D_{12}$ and $D_{43}$ can be considered the object is moved with same in-plane displacements $u$ and $v$ but measured at different locations in front of the camera. However, as shown in Table 11.1, the u-displacements for $D_{12}$ and $D_{43}$ are different; obviously, the $u$−displacement gradient along x-axis of $D_{12}$ is relatively larger than $D_{43}$; meanwhile, the DIC determined displacement contours for $D_{13}$, $D_{23}$, $D_{42}$ and $D_{43}$ are parallel to the vertical axis indicate that the pseudo in-plane displacement due to equilivent out-of-plane displacement dominates the in-plane displacement fields much more than the displacement introduced by physically moving the aluminum plate 3 mm to the right.

Ideally the DIC determined $u$ displacement field by using images taken at 102 mm focal length should be flat and the $v$ displacement should be vanished because the object is moved along x-axis with 3 mm nominal displacement; the determined mean values for u- and v-displacement are (3.0472 mm, 0.0952 mm) with standard derivation (0.0184 mm, 0.0172 mm). The DIC determined displacement is 1.57% deviated from the nominal displacement. However, as shown in Fig. 11.8a, the determined $u$ and $v$ displacement fields are both rotated with respect to y-axis and x-axis; similarly, the total displacement is also inclined with respect to y-axis; the inclination might be introduced by DIC algorithm or lens distortion; the displacement

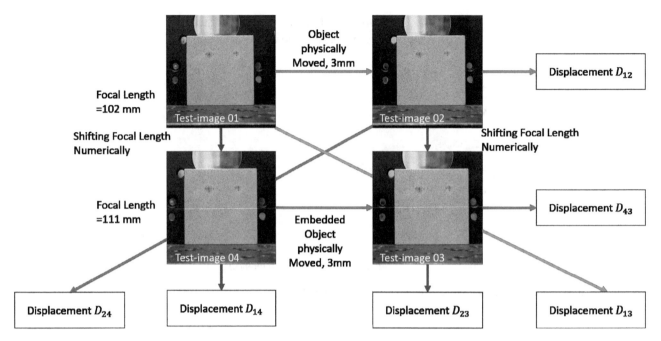

**Fig. 11.7** Possible displacement fields determined by DIC from an image-pair with physical movement and the corresponding computer generated image-pair with focal length numerically changed

**Table 11.1** Displacement components

| Displ. no. | $D_{12}$ | $D_{13}$ | $D_{14}$ | $D_{23}$ | $D_{42}$ | $D_{43}$ |
|---|---|---|---|---|---|---|
| In-plane displ. | +3 mm | +3 mm | – | – | +3 mm | +3 mm |
| Pseudo out-of-plane disp. | – | $\Delta z$ | $\Delta z$ | $\Delta z$ | $-\Delta z$ | – |
| $u$ – displacement (note: $D_{42} = -D_{24}$) | | | | | | |

fields are similar to the pseudo in-plane displacements determined by 2D DIC with images before and after focal length numerically adjusted. Based on this observation, in this study, images numerically generated by adjusting focal length are also applied for displacement determining to eliminate or reduce the inclinations of displacement fields.

In this study, image-pair obtained at focal length 102 mm before and after the aluminum plate is nominally moved +3 mm to the right, and then the corresponding image-pair with focal length adjusted to 111 mm are numerically generated. As indicated in Fig. 11.7, all independent displacement field $D_{ij}$ are calculated by VIC-2D. From Table 11.1, the displacement contours of $D_{13}$, $D_{23}$, $D_{42}$ and $D_{43}$ are similar but the embedded displacement fields are different; in this study, a linear combination of those four displacement fields are considered, the final displacement $u_i$ ($u_1 = u$, $u_2 = v$) is evaluated by following equation;

$$u_i = \left.\frac{D_{13} - D_{14} + D_{23} - D_{24}}{2}\right|_i , i = 1, 2 \qquad (11.1)$$

As shown in Fig. 11.8b, the $u$, $v$ and total displacement are no more inclined with respect to $y$-axis, $x$-axis and $y$-axis respectively; in addition, $u$, and total displacement fields become shallow-bowl in shape and $v$–displacement looks like to be twisted. The mean displacements ($\bar{u}, \bar{v}$) determined by Eq. (11.1) is (3.2504 mm, −0.0201 mm). It worth to notice that because the scale used for plotting displacement field is small with respect to the one used for Fig. 11.8a; therefor, the actual displacement deviations shown in Fig. 11.8b are not larger than the displacement deviations of Fig. 11.8a which are determined by the image-pair captured at 102 mm focal length. In fact, the corresponding standard deviations of $u$, $v$ displacement determined by using image-pairs both before and after adjusting the focal length are (0.0066 mm, 0.0069 mm) which are 1/3 of the displacement deviations determined by typical DIC method. Regarding the mean displacement value, as indicated in Table 11.2, the minimum $u$–displacement difference is given by $D_{12}|_u$, and $D_{43}|_u$ gives the maximum

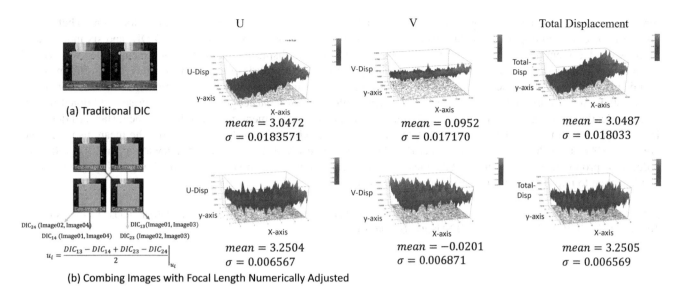

**Fig. 11.8** Displacement fields determined by combining $D_{ij}$

**Table 11.2** Mean and STD of the determined in-plan displacements

| Parameters | Displacement determined by | $u$ (mm) | $v$ (mm) | Total displ.(mm) |
|---|---|---|---|---|
| Mean | DIC(Image01, Image02), $D_{12}$ | 3.0472 | 0.0952 | 3.0487 |
|  | DIC(Image04, Image03), $D_{43}$ | 3.2514 | −0.0203 | 3.2514 |
|  | $u_i = \left. \frac{D_{13}-D_{14}+D_{23}-D_{24}}{2} \right|_i$ | 3.2504 | −0.0201 | 3.2505 |
| $\sigma$ | DIC(Image01, Image02), $D_{12}$ | 0.0184 | 0.0172 | 0.0181 |
|  | DIC(Image04, Image03), $D_{43}$ | 0.0075 | 0.0065 | 0.0075 |
|  | $u_i = \left. \frac{D_{13}-D_{14}+D_{23}-D_{24}}{2} \right|_i$ | 0.0066 | 0.0069 | 0.0066 |

$u-$displacement difference. As for $v-$displacement field, the minimum mean value in magnitude are given by $D_{43}|_v$ and displacement determined by proposed linear combination method, and $D_{12}|_v$ gives the maximum $v$-displacement difference. Considering Eq. (11.1) is linearly combination of DIC determined displacement fields; therefore, the corresponding focal length of the calculated displacement should be bounded within (102 mm, 111 mm); in this study, based on the aluminum plate is moved +3 mm to the right, it might can conclude that the mean $u-$displacements are increased as the focal length become longer or the object is "virtually" moved toward to the camera. As for mean $v-$displacements, similar phenomenon can be observed but the mean $v-$displacements are decreased as focal length increased (or the object is "virtually" moved toward to the camera). Meanwhile, according to the value of mean displacement given in Table 11.2, the mean $u-$displacement is about 6.7% larger than $D_{12}|_u$, and 8.4% larger with respect to nominal displacement.

## Conclusions

In this study, applying plenoptic camera with DIC method are explored; based on the experimental results and discussions, numerically changing the focal length of an image taken by using plenoptic camera can be analogized as to move the object toward or away from camera. Therefore, pseudo in-plane displacement and image size change can be calculated by using 2D DIC with images before and after numerically adjusting the focal length. The result showed that the image size is linearly enlarged as focal length increased, the magnitude of the pseudo in-plane displacement is symmetry with respect to the intersection point of optical axis and the plate, and the locations of image-center of the aluminum plate are slightly changed after focal length adjusted but no correlation with $\Delta f$ can be found.

For a given $\Delta f$, the focal length adjusting function is also used to numerically generate two additional images based on images captured at given focal length before and after test object moved in-plane. For a given $\Delta f$, the focal length adjusting function is also used to numerically generate two additional images based on images captured at given focal length before and after test object moved in-plane. The "displacement" of any two images are then evaluated by 2D DIC and six independent in-plane displacement fields can be determined; among all independent displacement fields, four are embedded with virtual out-of-plane displacement introduced by focal length changes. A simple formula is proposed in this study; by substituting those four displacement fields into the formula and calculates pixel by pixel and the modified displacement fields are then obtained. In this study, images are first taken by plenoptic camera with 102 mm focal length and then generates additional images by numerically adjusting the focal length to 111 mm, the modified displacement gives 1/3 displacement standard deviation in comparison with the one determined by original images (taken with 102 mm focal length) but the mean displacements are about 10% larger than the original results. Based on the current result, the displacement deviation can be significantly reduced by proposed simple formula but the mean displacement is increased, for next phase, weighting will be given to each displacement to improve the mean displacement determination.

**Acknowledgements** The partial financial support provided by the Ministry of Science and Technology of Taiwan, R.O.C (Award Numbers: 106-2221-E-492-013- and 107-2221-E-492-012-) is greatly appreciated.

# Chapter 12
# Identification of Interparticle Contacts in Granular Media Using Mechanoluminescent Material

Pawarut Jongchansitto, Damien Boyer, Itthichai Preechawuttipong, and Xavier Balandraud

**Abstract** Mechanoluminescent powders are new materials that can be considered as intelligent, active or responsive because they have the property of emitting light when they are mechanically deformed. They open perspectives for the measurement of stresses in mechanical parts. The present study focused on the stress concentrations in granular materials. Granular systems are defined as a collection of particles whose macroscopic behavior depends on the contact forces at the local scale. Some techniques are available for measurements in the bulk, such as X-ray tomography combined with volumetric digital image correlation. An extensive literature also deals with two-dimensional approaches: optical photography combined with digital image correlation, photoelasticimetry, infrared thermography. Mechanoluminescent materials offer new possibilities for revealing contact force networks in granular materials. Epoxy resin and mechanoluminescent powder were mixed to prepare dumbbell-like specimens and cylinders. Dumbbell-like specimens were used for preliminary uniaxial tensile tests. Cylinders were used to prepare granular systems for confined compression tests. Homogeneous light emission was obtained in the former case, while light concentrations were evidenced in the latter case.

**Keywords** Mechanoluminescence · Luminescence · Granular material · Interparticle contact

## Introduction

Mechanoluminescence is a physical phenomenon observed in certain crystals that are able to emit visible light when they are subjected to mechanical stimuli such as grinding, cutting, striking or friction [1, 2]. Mechanoluminescent (ML) materials can be used as sensors for structure monitoring [3–5]. Most of them are inorganic compounds constituted of an oxide host lattice doped with lanthanide ions [6–8] or more rarely with transition metal elements [9]. For example, $SrAl_2O_4$:($Eu^{2+}$,$Dy^{3+}$) exhibits strong mechanoluminescence [10, 11]. This phosphor emits green light upon mechanical stress and exhibits a long afterglow after photonic excitation in the ultraviolet-visible range [12]. The objective of the present study is to show the potentiality of this active substance to reveal interparticle contacts in granular materials.

A granular system is defined as a collection of particles whose macroscopic behavior depends on the contact forces at the local scale. Some techniques are available for measurements in the bulk, such as X-ray tomography combined with volumetric digital image correlation [13–15]. An extensive literature also deals with two-dimensional (2D) approaches: digital image correlation [16–22], photoelasticimetry [23–25], infrared thermography [26]. In the present study, epoxy resin and $SrAl_2O_4$:($Eu^{2+}$,$Dy^{3+}$) powder were mixed to prepare dumbbell-like specimens and cylinders. Dumbbell-like specimens were used for preliminary uniaxial tensile tests. Cylinders were used to prepare 2D cohesionless granular materials for confined compression tests.

## Material Elaboration and Experimental Setup

Figure 12.1a shows the SrAl$_2$O$_4$:(Eu$^{2+}$,Dy$^{3+}$) powder supplied by Honeywell. It is possible to excite the powder in the visible and ultraviolet ranges. The picture on the right shows the powder after excitation in the visible range, leading to the green color. The powder was incorporated into a two-component epoxy resin with a powder-to-epoxy mass ratio equal to 1:19. Figure 12.1b shows the mold used to prepare ML dumbbell-like specimens, 2.15 mm in thickness, 9.9 mm in width, with a useful length of 110 mm. Cylinders, 50 mm in length, were also prepared: see Fig. 12.1c. Note that the viscous epoxy-powder mixture was poured to a depth of 5 mm only, the remaining 45 mm were pure epoxy. Finally, curing was performed in an oven at 80 °C for 2 h. Figure 12.1d shows a dumbbell-like specimen and a cylinder after photonic excitation in the visible range.

Figure 12.2 presents the experimental setup for mechanical tests. Dumbbell-like specimens were placed in the grips of a ±20 kN Zwick testing machine and equipped with a clip-on extensometer, 50 mm in gauge length: see Fig. 12.2a. ML and non-ML (pure epoxy) specimens were tested for comparison purposes. Force-controlled loading was applied with a rate of 5 N/s up to 200 N for both types. Confined compression tests on two granular configurations were also performed using a rectangular metallic frame. They involved three ML cylinders (Fig. 12.2b) and fifteen cylinders among which three are ML (Fig. 12.2c). Displacement-controlled loading was applied with a rate of −0.02 mm/s: up to −500 N and −2000 N for the configurations with three and fifteen cylinders, respectively.

A CMOS Fuji X-T20 camera was used to capture images during mechanical loading: see Fig. 12.2c. Images were processed using Matlab. Each image was converted to greyscale, leading to pixel values of between 0 (black) and 1 (white). The analysis

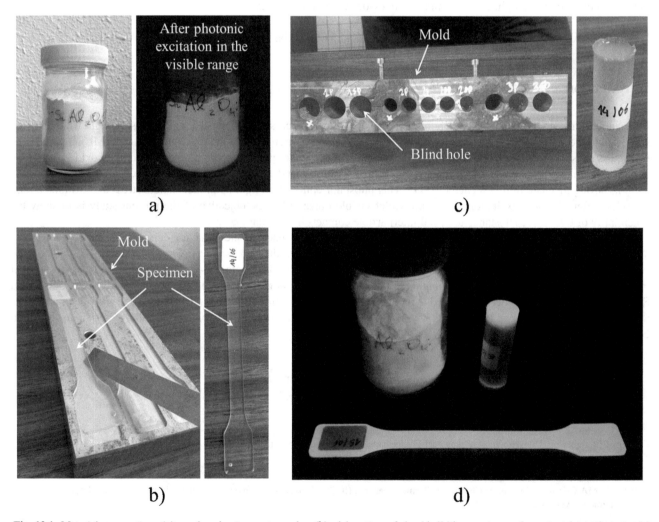

**Fig. 12.1** Material preparation: (**a**) mechanoluminescent powder, (**b**) elaboration of dumbbell-like specimens for uniaxial tensile tests, (**c**) elaboration of cylinders to be used for granular configurations, (**d**) dumbbell-like specimen and cylinder after photonic excitation in the visible range

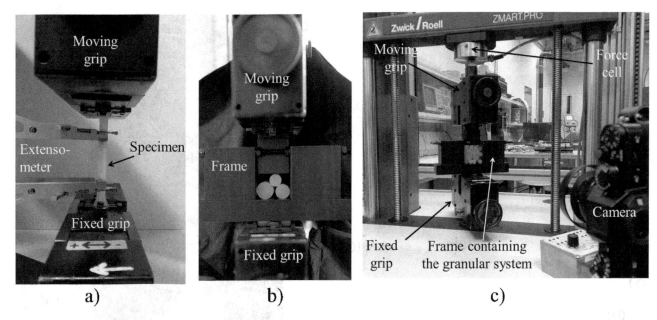

**Fig. 12.2** Experimental setup: (**a**) uniaxial tensile test on longitudinal specimen, (**b**) confined compression of three ML cylinders, (**c**) photo of an experiment involving fifteen cylinders among which three are ML

is then based on the change in light intensity with respect to an image captured just before starting the mechanical loading. Note that all the experiments were performed in the dark with a waiting time of 30 min before starting any measurement.

## Preliminary Uniaxial Tensile Tests

Figure 12.3a shows the stress-strain curves for the ML and non-ML dumbbell-like specimens. Stiffness is higher for the ML type. The ML powder acted as a mechanical reinforcement. Figure 12.3b shows the variation in the average light emission in the region of interest, corresponding to the zone between the two parts of the extensometer (see Fig. 12.3c). The light emission is not linear with respect to the applied force. For smaller forces, light emission is very low. Light emission increases progressively and tends to be linear from about 120 N. Figure 12.3c shows the field of light emission in the region of interest, revealing a quite homogenous distribution as expected for a homogeneous tensile test.

Other tests were perfected until specimen fracture. Fractography analysis was then performed by scanning electron microscopy (SEM) using secondary electrons. Figure 12.4 shows the topography of the specimen's fracture surface. Brittle failure mechanisms are visible: smooth fracture surfaces with river lines with the same orientation. ML grains are visible. Most of them are at the extremity of the lines with same orientation, meaning that they are probably fracture initiation zones. Magnifications revealed matrix/grain debonding due to the large difference in stiffness between the two components.

## Application to Granular Configurations

Figure 12.5a shows the distribution of light emission in the three ML cylinders subjected to a vertical compression force of −500 N. The light emission at the contact between the small and medium cylinders appears greater than the light emission at the contact between the small and large cylinders. This result can be justified by the orientation of the contact normal with respect to the vertical applied force. It can be also noted that the stress flow is more clearly evidenced in the smallest cylinder. Figure 12.5b shows the distribution of light emission in the second granular configuration (15 cylinders among which three ML cylinders) for a vertical force of −2000 N. Although the top ML cylinder was a priori in contact with six cylinders, it was actually subjected to three "active" contacts, in the sense that only three contacts actually transfer significant forces. This can be explained by the orientation of the contact normals with respect with the vertical applied force and by a heterogeneous distribution of the contact forces from the horizontal part. As a general comment, mechanoluminescence appears to provide

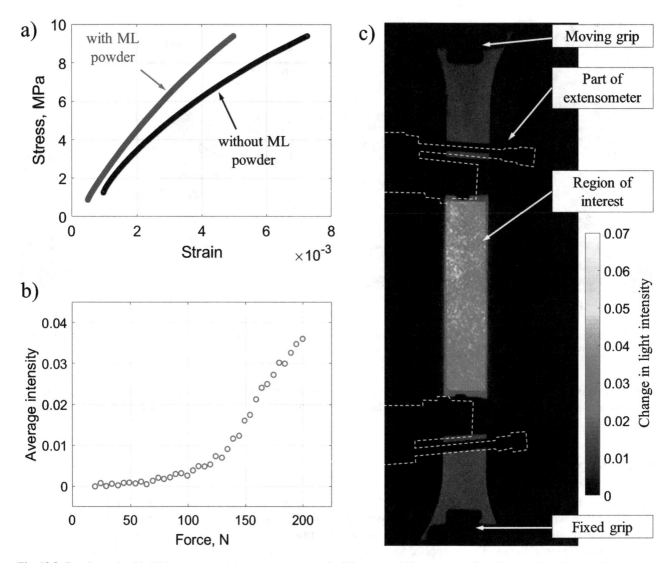

**Fig. 12.3** Results on dumbbell-like specimen: (**a**) stress-strain curves for ML and non-ML specimens, (**b**) average light emission in the region of interest as a function of the applied force, (**c**) distribution of light emission in the region of interest

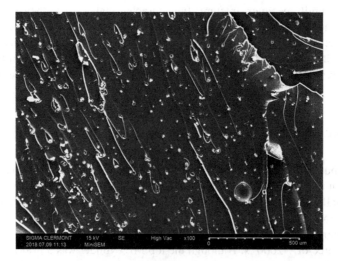

**Fig. 12.4** Fractography analysis by SEM of a ML dumbbell-like specimen

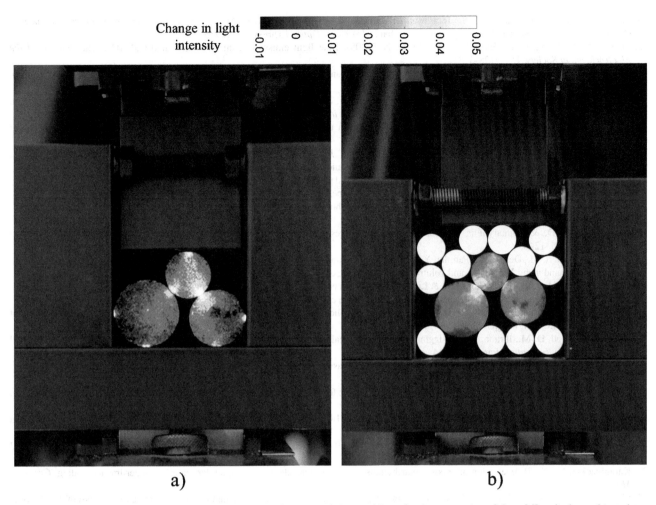

**Fig. 12.5** Light emission in granular systems, superimposed on optical photos: (**a**) confined compression of three ML cylinders subjected to a vertical force of −500 N, (**b**) confined compression of fifteen cylinders among which three are ML, under a vertical force of −2000 N

information about contact stress intensity between the cylinders. ML materials could be used for the micromechanical analysis of granular systems by revealing the contact network supporting the main part of the externally applied loading.

**Acknowledgements** The authors gratefully acknowledge the Ministère de l'Europe et des Affaires Etrangères (MEAE) and the Ministère de l'Enseignement supérieur, de la Recherche et de l'Innovation (MESRI) in France, as well as the Office of the Higher Education Commission (OHEC) of the Ministry of Education in Thailand. The authors also gratefully thank the French Embassy in Thailand and Campus France for their support during this research (PHC SIAM 2018, Project 40710SE). The authors would also like to acknowledge the financial support through the Research Grant for New Scholar (MRG6080251) from the Thailand Research Fund (TRF) and Thailand's Office of the Higher Education Commission (OHEC). Finally, the authors gratefully thank Mr. Maël Tissier, Sigma-Clermont Engineering School, for the elaboration of mechanoluminescent materials, as well as Mr. Clément Weigel and Mr. Alexis Gravier, Sigma-Clermont Engineering School, for the manufacturing of the testing device and the molds.

# References

1. Bünzli, J. C. G., & Wong, K. L. (2018). Lanthanide mechanoluminescence. *Journal of Rare Earth, 36*, 1–41.
2. Feng, A., & Smet, P. F. (2018). A review of mechanoluminescence in inorganic solids: compounds, mechanisms, models and applications. *Materials, 11*, 484.
3. Kamimura, S., Yamada, H., & Xu, C. N. (2012). Development of new elasticoluminescent material SrMg2(PO4)2:Eu. *Journal of Luminescence, 132*, 526–530.
4. Zhang, J. C., Xu, C. N., & Long, Y. Z. (2013). Elastico-mechanoluminescence in CaZr(PO4)2:Eu2+ with multiple trap levels. *Optics Express, 21*, 13699–13709.

5. Zhang, J. C., Xu, C. N., Kamimura, S., Terasawa, Y., Yamada, H., & Wang, X. (2013). An intense elastico-mechanoluminescence material CaZnOS:Mn2+ for sensing and imaging multiple mechanical stresses. *Optics Express, 21*, 12976–12986.
6. Zhang, H. W., Yamada, H., Terasaki, N., & Xu, C. N. (2008). Blue light emission from stress-activated CaYAl3O7:Eu. *Journal of the Electrochemical Society, 155*, J128–J131.
7. Zhang, H., Terasaki, N., Yamada, H., & Xu, C. N. (2009). Mechanoluminescence of Europium-doped SrAMgSi(2)O(7) (A = Ca, Sr, Ba). *Japanese Journal of Applied Physics, 48*, 04C109.
8. Zhang, J. C., Fan, X. H., Yan, X., Xia, F., Kong, W. J., Long, Y. Z., & Wang, X. S. (2018). Sacrificing trap density to achieve short-delay and high-contrast mechanoluminescence for stress imaging. *Acta Materialia, 152*, 148–154.
9. Tu, D., Xu, C. N., Fujio, Y., & Yoshida, A. (2015). Mechanism of mechanical quenching and mechanoluminescence in phosphorescent CaZnOS:Cu. *Light-Science & Applications, 4*, e356.
10. Yun, G. J., Rahimi, M. R., Gandomi, A. H., Lim, G. C., & Choi, J. S. (2013). Stress sensing performance using mechanoluminescence of SrAl2O4:Eu (SAOE) and SrAl2O4:Eu, Dy (SAOED) under mechanical loadings. *Smart Materials and Structures, 22*, 055006.
11. Yang, Y., Zheng, S. H., Fu, X. Y., & Zhang, H. W. (2018). Remote and portable mechanoluminescence sensing system based on a SrAl2O4:Eu,Dy film and its potential application to monitoring the safety of gas pipelines. *Optik, 158*, 602–609.
12. Li, Y., Gecevicius, M., & Qiu, J. R. (2016). Long phosphorescent phosphors-from fundamentals to applications. *Chemical Society Reviews, 45*, 2090–2136.
13. Wolf, H., Konig, D., & Triantafyllidis, T. (2003). Experimental investigation of shear band patterns in granular material. *Journal of Structural Geology, 25*, 1229–1240.
14. Hall, S. A., Bornert, M., Desrues, J., Pannier, Y., Lenoir, N., Viggiani, G., & Besuelle, P. (2010). Discrete and continuum analysis of localised deformation in sand using X-ray μCT and volumetric digital image correlation. *Geotechnique, 60*, 315–322.
15. Hu, Z. X., Du, Y. J., Luo, H. Y., Zhong, B., & Lu, H. B. (2014). Internal deformation measurement and force chain characterization of mason sand under confined compression using incremental digital volume correlation. *Experimental Mechanics, 54*, 1575–1586.
16. Slominski, C., Niedostatkiewicz, M., & Tejchman, J. (2007). Application of particle image velocimetry (PIV) for deformation measurement during granular silo flow. *Powder Technology, 173*, 1–18.
17. Hall, S. A., Wood, D. M., Ibraim, E., & Viggiani, G. (2010). Localised deformation patterning in 2D granular materials revealed by digital image correlation. *Granular Matter, 12*, 1–14.
18. Richefeu, V., Combe, G., & Viggiani, G. (2012). An experimental assessment of displacement fluctuations in a 2D granular material subjected to shear. *Geotechnique Letter, 2*, 113–118.
19. Marteau, E., & Andrade, J. E. (2017). A novel experimental device for investigating the multiscale behavior of granular materials under shear. *Granular Matter, 19*, 77.
20. Hurley, R., Marteau, E., Ravichandran, G., & Andrade, J. E. (2014). Extracting inter-particle forces in opaque granular materials: Beyond photoelasticity. *Journal of the Mechanics and Physics of Solids, 63*, 154–166.
21. Hurley, R. C., Lim, K. W., Ravichandran, G., & Andrade, J. E. (2016). Dynamic inter-particle force inference in granular materials: method and application. *Experimental Mechanics, 56*, 217–229.
22. Karanjgaokar, N. (2017). Evaluation of energy contributions using inter-particle forces in granular materials under impact loading. *Granular Matter, 19*, 36.
23. Shukla, A., & Damania, C. (1987). Experimental investigation of wave velocity and dynamic contact stresses in an assembly of disks. *Experimental Mechanics, 27*, 268–281.
24. Roessig, K. M., Foster, J. C., & Bardenhagen, S. G. (2002). Dynamic stress chain formation in a two-dimensional particle bed. *Experimental Mechanics, 42*, 329–337.
25. Mirbagheri, S. A., Ceniceros, E., Jabbarzadeh, M., McCormick, Z., & Fu, H. C. (2015). Sensitively photoelastic biocompatible gelatin spheres for investigation of locomotion in granular media. *Experimental Mechanics, 55*, 427–438.
26. Jongchansitto, P., Balandraud, X., Preechawuttipong, I., Le Cam, J. B., & Garnier, P. (2018). Thermoelastic couplings and interparticle friction evidenced by infrared thermography in granular materials. *Experimental Mechanics, 58*, 1469–1478.

# Chapter 13
# Colour Transfer in Twelve Fringe Photoelasticity (TFP)

**Sachin Sasikumar and K. Ramesh**

**Abstract** Colour transfer methodology is used in the field of image processing to transfer colour characteristics between images. In this study, colour transfer employing principal component analysis (PCA) is explored to circumvent the colour mismatch between application and calibration images in twelve fringe photoelasticity. For this, the images are considered as 3-D pixel clouds in *RGB* colour space.

**Keywords** Digital photoelasticity · TFP · Colour adaptation · Colour transfer · Principal component analysis

## Introduction

In TFP [1], a single dark field isochromatic image of the application specimen captured under white light is made use of in its fringe order demodulation. For this, a calibration table is generated first. The fringe order at any point in the application image is allotted by comparing its $R$, $G$, $B$ intensity values with that of the calibration table in a least squares sense. Fringe demodulation becomes erroneous if there is colour mismatch/tint variation between application and calibration images due to illumination source, variation in ambient lighting, annealing and stress freezing of the specimen, optical elements, camera characteristics [2]. Colour adaptation schemes were introduced to tackle the issue of colour mismatch.

One-point colour adaptation proposed by Madhu et al. [3] used zero load bright field images of the application and calibration specimens to effect adaptation. Neethi Simon and Ramesh made use of a single isochromatic image of the application specimen to modify the calibration table in two-point colour adaptation [4]. Two-point adaptation is a linear interpolation scheme which uses minimum and maximum colour channel intensities of application and calibration images as the interpolating parameters. A quadratic interpolation scheme that used mean value of pixel intensities in addition to minima and maxima was put forth by Swain et al. [5] which was christened as three-point colour adaptation.

In colour transfer techniques [6], which are widely used in image processing, colour characteristics are transferred from one image (*source image*) to another (*target image*). By considering application specimen image as the source and the calibration specimen image as the target, the utility of colour transfer in tackling colour mismatch is explored in this paper.

The usage of $\langle l, \alpha, \beta \rangle$ colour space [7] where a colour is expressed in terms of three numerical values, $\langle l \rangle$ for lightness, $\langle \alpha \rangle$ and $\langle \beta \rangle$ for (green—red) and (blue—yellow) colour components was opted by Reinhard et al. [6] for colour transfer between natural scenes. Histogram matching was also recommended as an additional step to achieve a complete transfer of distribution of images [8]. A comprehensive study on the suitable choice of colour spaces that can be used for colour transfer was done by Reinhard and Tania [9]. Recognizing the impracticality in devising a colour space that would work globally for all colour transfer problems, some authors have suggested to apply principal component analysis (PCA) [10] or independent Component analysis [11] to derive colour spaces that are specific to individual images.

The aim of colour transfer is to match the pixel distribution of the target image with that of the source image via geometrical transformation steps *viz.,* translation, scaling, rotation [12]. In the colour transfer proposed by Zhang et al. [12], mean of pixel intensities was used as translating parameter. Scaling parameter and rotation matrix were derived by performing singular value decomposition (SVD) on the covariance matrix of colour channel intensities. Following their work, translation of pixel clouds is decided by the mean, scaling factor and orientation is explored by principal component analysis (PCA) of covariance matrix of colour channel intensities.

S. Sasikumar · K. Ramesh (✉)
Department of Applied Mechanics, Indian Institute of Technology Madras, Chennai, India
e-mail: kramesh@iitm.ac.in

## Algorithm for Colour Transfer Using Principal Component Analysis (PCA)

Let $I^{cal}$ and $I^{app}$ denotes the pixel intensities of the calibration and application images.

$$I^{cal} = \begin{pmatrix} R^{cal}_{(x_1,y_1)} & R^{cal}_{(x_i,y_i)} \\ G^{cal}_{(x_1,y_1)} \vdots G^{cal}_{(x_i,y_i)} \vdots \\ B^{cal}_{(x_1,y_1)} & B^{cal}_{(x_i,y_i)} \\ 1 & 1 \end{pmatrix}; \quad I^{app} = \begin{pmatrix} R^{app}_{(x_1,y_1)} & R^{app}_{(x_i,y_i)} \\ G^{app}_{(x_1,y_1)} \vdots G^{app}_{(x_i,y_i)} \vdots \\ B^{app}_{(x_1,y_1)} & B^{app}_{(x_i,y_i)} \\ 1 & 1 \end{pmatrix}$$

In this section, superscripts *'cal'* and *'app'* denotes calibration and application images respectively. Subscript *'CT'* refers to colour transferred calibration image.

The algorithm consists of five steps—in step 1, the calibration image pixel cloud is translated with respect to its mean which in step 2 is rotated such that its dominant orientation gets aligned with the *RGB* coordinates, step 3 scales the calibration image pixel cloud, rotation in step 4 aligns the calibration image cloud along dominant orientation of the application image pixel cloud, in step 5 calibration image cloud is translated with respect to the mean of application image cloud. At the end of colour transfer, calibration image cloud will share the same mean, spread and orientation as that of the application image cloud. The five steps can be summarised as,

$$I^{cal}_{CT} = \left( T^{app} \cdot Z^{app} \cdot S^{app} \cdot S^{cal} \cdot Z^{cal} \cdot T^{cal} \right) \cdot I^{cal} \tag{13.1}$$

where *T, S, Z* denotes translation, scaling and rotation matrices respectively.

Let the means of *R, G, B* intensities of the application image be ($R^{app}_{Mean}$, $G^{app}_{Mean}$, $B^{app}_{Mean}$) and the calibration image be ($R^{cal}_{Mean}$, $G^{cal}_{Mean}$, $B^{cal}_{Mean}$), then the translation matrices can be represented as:

$$T^{app} = \begin{pmatrix} 1 & 0 & 0 & R^{app}_{Mean} \\ 0 & 1 & 0 & G^{app}_{Mean} \\ 0 & 0 & 1 & B^{app}_{Mean} \\ 0 & 0 & 0 & 1 \end{pmatrix} \quad \text{and} \quad T^{cal} = \begin{pmatrix} 1 & 0 & 0 & -R^{cal}_{Mean} \\ 0 & 1 & 0 & -G^{cal}_{Mean} \\ 0 & 0 & 1 & -B^{cal}_{Mean} \\ 0 & 0 & 0 & 1 \end{pmatrix} \tag{13.2}$$

Scaling factor and rotation matrices are computed by applying principal component analysis (PCA) on $I^{cal}$ and $I^{app}$ using the inbuilt function *pca*( ) available in Matlab®. The function *pca*( ) returns three outputs for a given input dataset *K*,

$$[\text{coefficient}, \text{ score}, \text{ latent}] = \text{pca}(K);$$

The *coefficient* matrix will be the rotation matrix for the input data, since it contains the eigenvectors of the covariance matrix of *K*. If $C^{cal}$ and $C^{app}$ denote the coefficient matrices of calibration and application images, the rotation matrices $Z^{cal}$ and $Z^{app}$ can be written as:

$$Z^{cal} = \left( C^{cal} \right)^{-1}, Z^{app} = C^{app} \tag{13.3}$$

*Latent* vector contains the eigenvalues corresponding to the eigenvectors in the coefficient matrix. Let $\{\lambda^{cal}_1, \lambda^{cal}_2, \lambda^{cal}_3\}^T$ and $\{\lambda^{app}_1, \lambda^{app}_2, \lambda^{app}_3\}^T$ be the latent vectors for calibration and application images, then the scaling matrices are given as:

$$S^{cal} = \text{diag}\left(\sqrt{\lambda^{cal}_1}, \sqrt{\lambda^{cal}_2}, \sqrt{\lambda^{cal}_3}, 1\right) \quad \text{and} \quad S^{app} = \text{diag}\left(\sqrt{\lambda^{app}_1}, \sqrt{\lambda^{app}_2}, \sqrt{\lambda^{app}_3}, 1\right) \tag{13.4}$$

## Performance of the Proposed Algorithm

Circular disc under diametral compression (Fig. 13.1a) made of epoxy (diameter = 60 mm, load = 519.13 N, $F_\sigma = 12.15$ N/mm/fringe) is considered as the application specimen. The calibration specimen shown in Fig. 13.1b is also made of epoxy but has a tint variation from the application specimen due to aging effect. Both the images are captured using DSLR camera (Canon 450 D) under discrete fluorescent lamp (Philips-MASTER PL-C 18W/827/2P 1CT/5X10BOX). Since the calibration table is prepared by allotting integer fringe order values to the minima of green channel intensity [1], a 0–4 calibration table can be prepared from Fig. 13.1b. The pixel distribution in *RGB* colour space corresponding to Fig. 13.1a, b is represented in Fig. 13.1d, e respectively. The principal components marked along with the pixel distribution shows that the application and calibration images have different dominant orientations in the colour space. Figure 13.1c shows the calibration image in Fig. 13.1b after the application of colour transfer algorithm. The pixel distribution of the colour transferred calibration image is shown in Fig. 13.1f along with its principal components. Note that, after colour transfer the principal components of application and calibration images have become identical. The superimposed pixel distribution of the images before colour transfer (Fig. 13.2a) and after colour transfer (Fig. 13.2b) also brings out the potential of the method in matching pixel distributions.

On scrutiny of the *RGB* variation of the colour transferred image (Fig. 13.1c), it is evident that the green channel modulation is lost after the third minima. The modulation of other colour channels have also changed. So, the algorithm has altered the inherent modulation of the original calibration image and the modified image can no longer be used to generate 0–4 calibration table. The change in modulation can be attributed to the rotation of pixel clouds involved in the current algorithm.

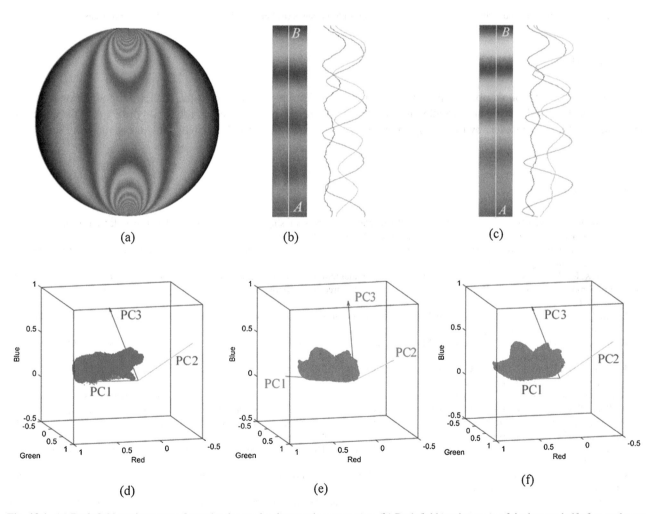

**Fig. 13.1** (a) Dark field isochromatic of circular disc under diametral compression, (b) Dark field isochromatic of the bottom half of central zone of C-specimen with its *R, G, B* intensity variation along line *AB*, (c) Image 'b' after colour transfer (with its *R, G, B* intensity variation along line *AB*), pixel distribution along with principal components corresponding to (d) Image 'a', (e) Image 'b', (f) Image 'c'

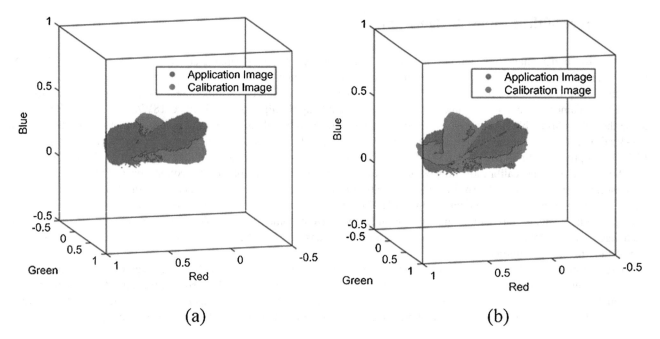

**Fig. 13.2** Superimposed pixel distribution of application and calibration images (**a**) before colour transfer, (**b**) after colour transfer

## Conclusion

It is seen that the rotation of pixel clouds involved in the algorithm has attributed to the change in modulation of colour channel intensities of the original calibration image. Hence, the study has shown the non-feasibility of PCA in the context of TFP. Further studies are underway to develop an improved colour transfer for TFP involving only translation and scaling.

## References

1. Ramesh, K., Ramakrishnan, V., & Ramya, C. (2015). New initiatives in single-colour image-based fringe order estimation in digital photoelasticity. *Journal of Strain Analysis for Engineering Design, 50*, 488–504.
2. Martínez-Verdú, F., Chorro, E., Perales, E., Vilaseca, M., & Pujo, J. (2010). 6—Camera-based colour measurement. In *Colour measurement* (p. 147-e2). Cambridge: Woodhead Publishing.
3. Madhu, K. R., Prasath, R. G. R., & Ramesh, K. (2007). Colour adaptation in three fringe photoelasticity. *Experimental Mechanics, 47*, 271–276.
4. Neethi Simon, B., & Ramesh, K. (2011). Colour adaptation in three fringe photoelasticity using a single image. *Experimental Techniques, 35*(5), 59–65.
5. Swain, D., Thomas, B. P., Philip, J., & Pillai, S. A. (2015). Novel calibration and color adaptation schemes in three-fringe RGB photoelasticity. *Optics and Lasers in Engineering, 66*, 320–329.
6. Reinhard, E., Ashikhmin, M., Gooch, B., & Shirley, P. (2001). Colour transfer between images. *IEEE Computer Graphics and Applications, 21*(5), 34–41.
7. Ruderman, D. L., Cronin, T. W., & Chiao, C. C. (1998). Statistics of cone responses to natural images: implications for visual coding. *Journal of the Optical Society of America A, 15*(8), 2036–2045.
8. Neumann, L., & Neumann, A. (2005). Color style transfer techniques using hue, lightness and saturation histogram matching. In L. Neumann, M. Sbert, B. Gooch, & W. Purgathofer (Eds.), *Proceedings of Computational aesthetics in graphics, visualization and imaging* (pp. 111–122).
9. Reinhard, E., & Pouli, T. (2011). Colour spaces for colour transfer. In R. Schettini, S. Tominaga, & A. Trémeau (Eds.), *Computational color imaging. CCIW, lecture notes in computer science* (Vol. 6626). Berlin: Springer.
10. Abadpour, A., & Kasaei, S. (2007). An efficient PCA-based color transfer method. *Journal of Visual Communication and Image Representation, 18*, 15–34.
11. Grundland, M., & Dodgson, N. (2005). *The decolorize algorithm for contrast enhancing, color to grayscale conversion.* Technical Report UCAM-CL-TR-649, University of Cambridge.
12. Zhang, X., Jiang, W., Liu, X., & Lei, X. (2018). Visualization of colour transfer between images in RGB colour space. In *International workshop on advanced image technology (IWAIT)*.

# Chapter 14
# Infrared Deflectometry

H. Toniuc and F. Pierron

**Abstract** This paper illustrate the use of deflectometry in the infrared spectrum to measure surface slopes on a plate deformed in bending.

**Keywords** Deflectometry · Infrared imaging · Plate bending · Virtual fields method

## Introduction

Deflectometry is a technique that measures surface slopes of specularly reflective objects. The motivation for this is mostly defect detection on shiny objects like painted car bodies [1]. Deflectometry requires a complex calibration process as detailed in [2]. In the experimental mechanics community, the technique is used to measure the slope deformation of plates loaded in bending. From the initial paper by Ligtenberg [3], the method has been used sporadically by a few groups worldwide [4–11], mostly for material characterization purposes. The main advantage of deflectometry is its very high sensitivity that can be tuned independently from the spatial resolution by tailoring the specimen to target distance. The slope resolution can reach down to 1 μrad [12, 13]. In [14], using an ultra-high speed camera, Lamb waves were measured at the surface of glass and composite panels, with peak to peak amplitudes as low as 20 nm. More recently, deflectometry coupled to the Virtual Fields Method was shown to provide spatially and temporally-resolved pressure distributions [15, 16]. It should be noted however that apart from [17, 18], all applications were on flat panels in normal incidence, most probably because of the more complex calibration and the lack of commercially available systems.

The main drawback of deflectometry however is that the surface under inspection needs to be specularly (or mirror-like) reflective. According to the Rayleigh criterion [19], a surface is predominantly specularly reflective if:

$$\frac{\lambda}{\sigma \cos \theta} > 8 \qquad (14.1)$$

where $\lambda$ is the wavelength of the light, $\sigma$ the surface RMS roughness and $\theta$ the light incidence angle from the surface normal. The relationship above shows clearly that by using a shallow incidence angle, specular reflection (SR) is enhanced. For normal incidence, in the visible spectrum ($\lambda \approx 500$ nm), SR is predominant for $\sigma < 60$ nm. Most as-manufactured engineering surfaces do not comply with this and coatings have been devised to make surfaces reflective [20]. However, imaging in the long infrared significantly relaxes this constraint. Indeed, at 12 μm, surfaces up to 1.5 μm RMS roughness are reflective at normal incidence.

To the best knowledge of the present authors, infrared (IR) deflectometry has first been reported in 2005 by Horbach and Kammel [21] but only in the 2010s has the technique started to attract more attention [22–25]. In the experimental mechanics community, a feasibility study has been recently published by the present authors [26] showing great potential for this technique. This paper will show an example of application on a bending test. Full details have been published in [26].

---

H. Toniuc · F. Pierron (✉)
Faculty of Engineering and Physical Sciences, University of Southampton, Southampton, UK
e-mail: ht1g15@soton.ac.uk; F.Pierron@soton.ac.uk

## Experimental Set-Up

The experimental set-up is shown in Fig. 14.1. The first element is a brushed aluminium plate, not reflective in the visible spectrum but so in the IR spectrum. This plate was held in a test frame to allow for bending deformation to be applied. The detailed bending configuration is shown on the left image in Fig. 14.2. The second element is the target grid. This was obtained by printing black squares onto a brushed aluminium plate. The difference in emissivity between the ink and the aluminium means that an IR pattern is generated when the plate is heated up, as in [21]. Here, the plate was heated with a hair dryer, up to about 50 °C. Finally, an infrared camera was used to image the reflection of the grid pattern onto the test specimen. A micro-bolometer array camera, FLIR FLIR A655SC 25°, was used to record the reflected images.

## Results

A reference image was taken before loading; loading was then applied through an in-house built system that incorporates load measurement (see right-hand side image of Fig. 14.2). The grid method algorithm presented in [27] was used to obtain spatial phases, which, after unwrapping, were changed into slopes using the sensitivity coefficient $p/4\pi h$ where h is the plate to grid distance and p is the grid pitch. Centred finite difference was then used to obtain curvatures (without any spatial

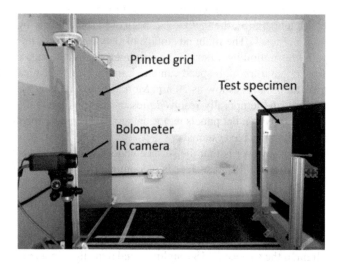

**Fig. 14.1** Experimental set-up

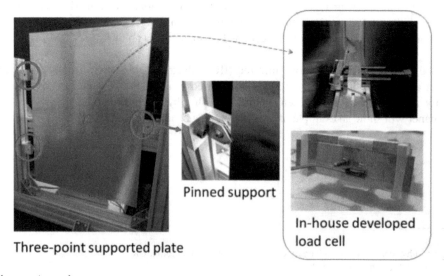

**Fig. 14.2** Details of the experimental set-up

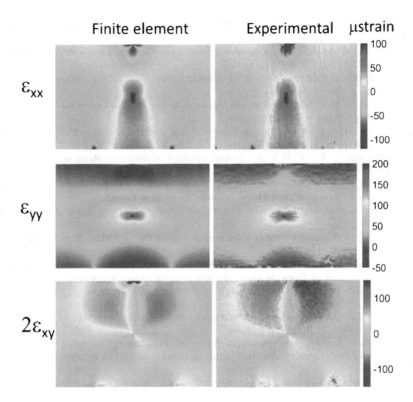

**Fig. 14.3** Comparison between finite element and experimental surface strains

smoothing) and finally, surface strains were recovered from curvatures by multiplying curvature by t/2 where t is the plate thickness (according to thin plate theory). Figure 14.3 shows a comparison between experiment and finite element model, demonstrating excellent correlation.

## Conclusion and Future Work

This paper demonstrates that it is feasible to measure the deformation of plates in bending with infrared deflectometry, something that, to the best knowledge of the authors, had not been achieved before. The rapid expansion of the technology of micro-bolometer array IR sensors opens-up the way for IR deflectometry. The next stage is to extend it to curved surfaces using the algorithm presented in [18].

**Acknowledgments** The authors are grateful to Dr. Devlin Hayduke from the Materials Sciences Corporation, Horsham, PA, USA for suggesting the idea, and to Dr. Yves Surrel for useful discussions on deflectometry and grid printing. Horea Toniuc acknowledges funding through the Excel Southampton Internship Programme.

## References

1. Arnal, L., et al. (2017). Detecting dings and dents on specular car body surfaces based on optical flow. *Journal of Manufacturing Systems, 45*, 306–321.
2. Balzer, J., & Werling, S. (2010). Principles of shape from specular reflection. *Measurement, 43*(10), 1305–1317.
3. Ligtenberg, F. K. (1954). A new experimental method for the determination of moments in small slab models. *Proceedings of SESA XII, 2*, 83–98.
4. Chiang, F. P., & Treiber, J. (1970). A note on Ligtenberg's reflective moiré method—Technical note offers an additional improvement to the optical arrangement whereby the sensitivity of the method can be changed at will. *Experimental Mechanics, 10*(9), 537–538.
5. Asundi, A. (1994). Novel techniques in reflection moiré. *Experimental Mechanics, 34*(3), 230–242.
6. Surrel, Y., et al. (1999). Phase-stepped deflectometry applied to shape measurement of bent plates. *Experimental Mechanics, 39*(1), 66–70.

7. Sciammarella, C. A., Trentadue, B., & Sciammarella, F. M. (2000). Measurement of bending stresses in shells of arbitrary shape using the reflection moire method. *Experimental Mechanics, 40*(3), 282–288.
8. Syed-Muhammad, K., et al. (2008). Characterization of composite plates using the virtual fields method with optimized loading conditions. *Composite Structures, 85*(1), 70–82.
9. Kim, J.-H., et al. (2009). Local stiffness reduction in impacted composite plates from full-field measurements. *Composites Part A: Applied Science and Manufacturing, 40*(12), 1961–1974.
10. Devivier, C., Pierron, F., & Wisnom, M. (2012). Damage detection in composite materials using full-field slope measurements. *Composites Part A: Applied Science and Manufacturing, 43*(10), 1650–1666.
11. Periasamy, C., & Tippur, H. V. (2013). A full-field reflection-mode digital gradient sensing method for measuring orthogonal slopes and curvatures of thin structures. *Measurement Science and Technology, 24*(2), 025202.
12. Devivier, C., Pierron, F., & Wisnom, M. R. (2013). Impact damage detection in composite plates using deflectometry and the Virtual Fields Method. *Composites Part A: Applied Science and Manufacturing, 48*, 201–218.
13. Devivier, C., Seghir, R., & Pierron, F. (2016). Deflectometry: full-field deformation measurements for composites NDT. In *Advanced structural health management and composite structures (ASHMCS)*, 2016, Jeonju, Republic of Korea.
14. Devivier, C., et al. (2016). Time-resolved full-field imaging of ultrasonic Lamb waves using deflectometry. *Experimental Mechanics, 56*(3), 345–357.
15. O'Donoughue, P., Robin, O., & Berry, A., (2018). Time-resolved identification of mechanical loadings on plates using the virtual fields method and deflectometry measurements. *Strain*, e12258–n/a.
16. Kaufmann, R., Ganapathisubramani, B., & Pierron, F. (2018). Full-field pressure reconstruction using deflectometry and the Virtual Fields Method. In *19th International Symposium on Applications of Laser and Imaging Techniques to Fluid Mechanics*.
17. Sciammarella, C. A., & Piroozan, P. (2007). Real-time determination of fringe pattern frequencies: An application to pressure measurement. *Optics and Lasers in Engineering, 45*(5), 565–577.
18. Surrel, Y., & Pierron, F. (Eds.). (2019). Deflectometry on curved surfaces. In *2018 SEM Conference ed. Conference Proceedings of the Society for Experimental Mechanics Series* (Vol. 3). Berlin: Springer.
19. Ogilvy, J. A. (1991). *Theory of wave scattering from random rough surfaces*. Boca Raton: CRC Press.
20. Kim, J.-H., et al. (2007). A procedure for producing reflective coatings on plates to be used for full-field slope measurements by a deflectometry technique. *Strain, 43*(2), 138–144.
21. Horbach, J. W., & Kammel, S. (2005). Deflectometric inspection of diffuse surfaces in the far-infrared spectrum. In: *Electronic imaging 2005*. SPIE.
22. Höfer, S., Burke, J., & Heizmann, M. (2016). Infrared deflectometry for the inspection of diffusely specular surfaces. *Advanced Optical Technologies, 5*(5–6), 377.
23. Sarosi, Z., et al. (2010). Detection of surface defects on sheet metal parts by using one-shot deflectometry in the infrared range. In: *InfraMation*, Las Vegas, Nevada.
24. Kim, D. W., et al. (2016). Extremely large freeform optics manufacturing and testing. In: *2015 11th Conference on Lasers and Electro-Optics Pacific Rim, CLEO-PR 2015*.
25. Su, T., et al. (2013). Measuring rough optical surfaces using scanning long-wave optical test system. 1. Principle and implementation. *Applied Optics, 52*(29), 7117–7126.
26. Toniuc, H., & Pierron, F. (2019). Infrared deflectometry for surface slope deformation measurements. *Experimental Mechanics*. Accepted.
27. Grédiac, M., Sur, F., & Blaysat, B. (2016). The Grid Method for in-plane displacement and strain measurement: a review and analysis. *Strain, 52*(3), 205–243.

# Chapter 15
# Real-Time Shadow Moiré Measurement by Two Light Sources

Fa-Yen Cheng, Terry Yuan-Fang Chen, Chia-Cheng Lee, and Ming-Tzer Lin

**Abstract** For accurate measurement, phase-shifting technique is usually adopted to the shadow moiré measurement system. Accurately introducing the amount of phase shift is required in order to extract the phase properly. However, the specimen or system may be moved during the time of image capture, and not suitable for real-time measurement. In order to overcome this drawback and make an in-line measurement, a shadow moiré system consisted of two light source of different colors and a color CCD camera is proposed. The phase shift is introduced by using two light sources illuminate the grating from different position simultaneously. The two moiré fringe patterns are captured by the color CCD camera, and are processed by a fringe analysis scheme using spiral phase transform (SPT) and optical flow techniques. The proposed fringe analysis scheme was applied to a simulated surface and a real specimen. The test results are reported and the validity of the scheme is investigated.

**Keywords** Shadow moiré · Real-time · Spiral phase transform · Optical flow

## Introduction

Shadow moiré method is an effective optical technique for surface profile measurement of diffusely reflecting objects. Due to cheap and easy to implement in industry environment, the method has been widely used in the semiconductor industry for distortion/warpage evaluation under thermal and/or mechanical loading. Phase-shifting technique is usually adopted to improve the sensitivity of shadow moiré measurement system. However, accurate introducing the amount of phase shift is required in order to extract the phase properly. If the specimen or system moved during the time of image capture, an error would be caused. Gómez-Pedrero et al. [1] rearrange the shadow moiré's set-up and use direct phase demodulation to calculate surface topography by a single fringe pattern. Du et al. [2] use similar set-up and spiral phase quadrature transform (SPT) method to calculate surface topography by two fringe patterns. In this paper a real-time shadow moiré measurement system, consisted of two light sources of different color and a color CCD camera was developed. In this system, the phase shift is introduced by using two light sources illuminate the grating from different position simultaneously. The two moiré fringe patterns are captured by a color CCD camera simultaneously, and separated into three monochromatic (red, green, and blue) fringe patterns for phase measurement. Employing spiral phase transform (SPT) and optical flow techniques, the unwrapped phase of one image can be determined to calculate the surface profile of object. Test of the system on a simulated specimen and a real specimen is reported.

## Principles

Figure 15.1 shows the schematic of experimental set-up for shadow moiré with two light sources. The intensity of moiré fringe pattern, I, can be described by

$$I(x, y) = a(x, y) + b(x, y) \cos\left[\theta(x, y)\right] \qquad (15.1)$$

---

F.-Y. Cheng · T. Y.-F. Chen (✉) · C.-C. Lee
Department of Mechanical Engineering, Nation Cheng Kung University, Tainan, Taiwan, Republic of China
e-mail: ctyf@mail.ncku.edu.tw

M.-T. Lin
Graduate Institute of Precision Engineering, National Chung Hsing University, Taichung, Taiwan
e-mail: mingtlin@dragon.nchu.edu.tw

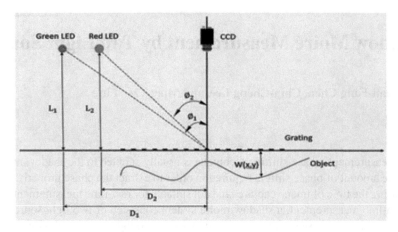

**Fig. 15.1** Schematic of set-up for shadow moiré with two light sources

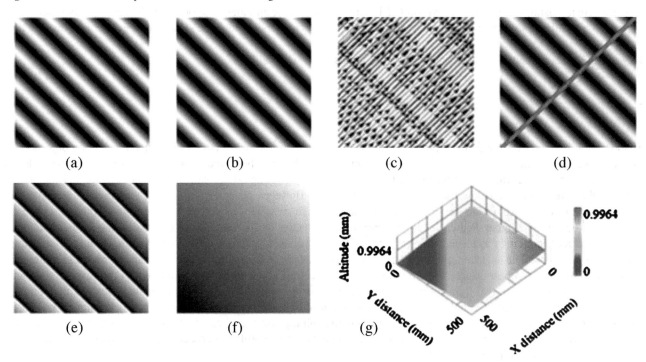

**Fig. 15.2** Simulated inclined plate test. Fringe patterns of incident angle, (**a**) 45° and (**b**) 43.5°, (**c**) fringe direction, (**d**) transformed sin θ, (**e**) wrapped phase, (**h**) unwrapped phased, and (**g**) reconstructed surface

where (x, y) is the position, a(x, y) is the background intensity, b(x, y) is the amplitude, and θ(x, y) is the phase of moiré fringe. After removing the DC term, SPT operator can be applied to obtained cosine signals and their quadrature signals given as [2]

$$b(x, y) \sin[\theta(x, y)] = -i \exp(-i\beta) \, \mathrm{SPT}\{\bar{I}(x, y)\} \tag{15.2}$$

The parameter, β, in Eq. 15.2 can be obtained by calculating the fringe orientation angle using optical flow approach [3]. The wrapped phase can be determined from

$$\theta(x, y) = \arctan\left\{\left[-i \exp(-i\beta) \, \mathrm{SPT}\{\bar{I}(x, y)\}\right] / \bar{I}(x, y)\right\} \tag{15.3}$$

After unwrapping, the height of surface can be calculated.

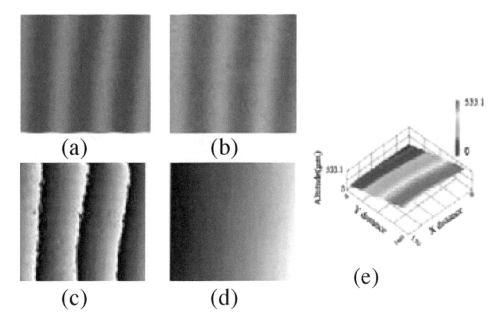

**Fig. 15.3** Inclined plate test. Fringe patterns of (**a**) incident angle 45° and (**b**) 43.15°, (**c**) wrapped phase, (**d**) unwrapped phase, (**e**) reconstructed surface

## Test Results and Conclusion

Moiré fringe patterns of an inclined flat plate was simulated with light incident angles of 45° and 43.5°, respectively. Test results are shown in Fig. 15.2. An error of 1.5% is achieved on the maximum height measurement of 1 mm. Figure 15.3 shows the real-time captured two images, wrapped and unwrapped phase maps, of an inclined plate specimen with light incident angles of 45° and 43.15°, respectively. Comparison to the result obtained by using four-step phase shifting method, an average error of 3.5 μm can be achieved between them for a height of 372 μm.

**Acknowledgements** The authors gratefully acknowledge the financial support provided to this study by the Ministry of Science and Technology of Taiwan under Grant No. MOST 106-2221-E-006 -106 -MY2.

## References

1. Gómez-Pedrero, J. A., Quiroga, J. A., José Terrón-López, M., & Crespo, D. (2006). Measurement of surface topography by RGB shadow-Moiré with direct phase demodulation. *Optics and Lasers in Engineering, 44*(12), 1297–1310.
2. Du, H., Zhao, H., & Li, B. (2013). Fast phase shifting shadow moiré by utilizing multiple light sources. *Proceeding of SPIE, 8681*, 86812A-6.
3. Vargas, J., Quiroga, J. A., Sorzano, C. O. S., Estrada, J. C., & Carazo, J. M. (2011). Two-step interferometry by a regularized optical flow algorithm. *Optics Letters, 36*(17), 3485–3487.

# Chapter 16
# Study of MRI Compatible Piezoelectric Motors by Finite Element Modeling and High-Speed Digital Holography

Paulo A. Carvalho, Haimi Tang, Payam Razavi, Koohyar Pooladvand, Westly C. Castro, Katie Y. Gandomi, Zhanyue Zhao, Christopher J. Nycz, Cosme Furlong, and Gregory S. Fischer

**Abstract** The use of intra-operative images with a magnetic resonance imager (MRI) can enable more precise surgical procedures when treating deep brain tumors. The constrained space inside the imager creates the need for robotic assistive devices. However, the strong static magnetic fields, fast changing magnetic gradients and sensitivity to electrical noise create challenges that can be partly addressed by a novel class of MRI compatible resonant piezoelectric motors to actuate the robots. These motors consist of a stator that is mechanically excited by a bonded piezoelectric ring and a frictionally coupled rotor. Steady-state excitation at certain frequencies leads to specific mode shapes with surface waves having both in- and out-of-plane displacement components. The interaction between the surface waves of the stator with the rotor results in the rotor spinning. Optimal operation of these motors depends on the stator's mode shapes generating waves that result in maximizing torquing of the rotor while reducing tangential force components. We present a finite element multi-physics model of the stator using COMSOL in frequency and time-domain. We combine the FEM with stroboscopic and time averaged, high-speed digital holography to create a validated model useful for optimizing motor design and performance. The methodology is used in the study of 30 mm diameter 40–60 kHz driven motor that have been demonstrated in an in-bore MRI compatible surgical robot.

**Keywords** Finite element modeling · Holographic measurements · MRI compatible actuation · Image guided surgery · Resonant piezoelectric motor

## Introduction

Cancer is the second leading cause of death in the United States estimated to have taken 600,920 lives in 2017 [1]. Cancer treatments have steadily advanced in the past decades and today includes immunotherapy [2], chemotherapy, radiation and surgical intervention. In the latter, being able to leave adequate margins is paramount to reduce remission [3]. An example of surgical intervention, is interstitial deep brain tumor ablation which requires real time imaging to guide the surgeon during the procedure. Today, most surgeries are performed based off of preoperative images registered to a patient with no ability to update the images during the procedure [4]. In some cases, a combination of a low resolution fast acquisition rate device such as ultrasound is used during the procedure to correct for target shift from previously acquired high resolution images [5]. The use of MRI images intra-operatively can address some of the limitations of a combination approach, better ensuring appropriate margins are achieved.

The strong magnetic fields of an MRI machine, fast changing gradients, its susceptibility to electrical noise and constrained space pose challenges to conducting surgical procedures within it. A robotic device is well suited to address these challenges. A key element of MRI compatible robots is their actuator. Some examples include using pneumatic [6–8], hydraulic [9, 10] and piezoelectric [11, 12] actuators for this application.

Piezoelectric resonant motors (PRM), also called ultrasonic motors, have been shown to work inside an MRI machine [12, 13]. Previous work has also demonstrated that these motors have the potential for reducing their distortion and noise in images by changing their construction [14]. In the aforementioned publication, the enclosure of the motor is modified to reduce metal content but the stator remains the same. Improving the stator would be the next step in improving MRI compatibility.

PRMs consist of two main components: A stator and a rotor. The rotor, usually cylindrical in shape, is spring loaded against the stator and constrained from translational motion. The stator, to which a piezoelectric ring is bonded, is responsible for inducing rotation of the rotor. The rotation is caused by exciting the piezoelectric ring to cause a desirable combined mode in the stator. The tangential motion of the surface at the peaks of the stator deformation during the induced mode are frictionally coupled to the rotor leading to its rotation. The fundamentals for the operation of PRMs is presented in [15].

The stator, being responsible for the mechanical deformations that lead to motor motion, determines most of the desired motor properties. Most MRI robots using commercial PRM motors tend to have 30 mm diameter stators, or more since torque increases with motor radii [15], and tend to operate with excitation signals of 40–60 KHz to induce deformations on the order of 1 $\mu$m. Whole field imaging of these stators during operation exceeds the imaging specifications of most visible light stereo camera systems. Digital holography can meet the needs for imaging these motors. In it, a single laser source is beam split into a coherent wavefront and a second wavefront that is shown on the object of interest and its reflected scattered light combined with the coherent wavefront. The interference pattern resultant from the interaction of the two wavefronts generates a hologram that can be digitally stored and processed to reveal sub wavelength features in the structure [16]. Combining this technique with dual exposure or strobing the laser allows recording of deformations in the structure that occur beyond the nominal frame rate of the camera.

In this work, we improve upon our previous finite element model (FEM) presented in [17] and test it experimentally through the use of digital holography. We use stroboscopic holography for measuring the time dependent evolution of the stator deformations and time-averaged holography to evaluate mode shapes.

## Simulation

A stator assembly based on the commercially available USR30 motor (Fukoku, Japan) was made in SolidWorks 2018 (Dassault Systèmes SE, France) and imported into COMSOL 5.3a (COMSOL Inc, Sweden) using the CAD import module. The assembly consists of the stator with 48 teeth, a thin epoxy layer and the piezoceramic crystal segmented into its electrically corresponding pieces. Figure 16.1a shows the dimensional drawings of the model.

The multi-physics simulation was based on [17]. Unlike the aforementioned simulation, the mesh size was specified as in Table 16.1 to account for the increased detail in the stator model. Resulting meshing can be seen in Fig. 16.1b. The solid mechanics fixed constraint was applied to the top and bottom of the indent in the center of the stator. This improved constraint

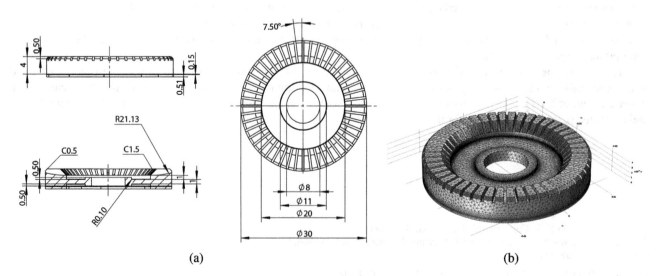

**Fig. 16.1** (a) Mechanical drawings for stator assembly. (b) Orthogonal view of meshed stator assembly in COMSOL

**Table 16.1** Parameters for tetrahedral meshing

| Parameter | Value |
|---|---|
| Max element size | 1.65e−3 |
| Min element size | 1.2e−4 |
| Max element growth rate | 1.4 |
| Curvature factor | 0.4 |
| Resolution narrow regions | 0.7 |

**Table 16.2** Basic mechanical properties of stator components

| Material | Density [kg/m$^3$] | Poisson's ratio | Young's modulus [GPa] |
|---|---|---|---|
| Brass | 8960 | 0.33 | 96 |
| Epoxy | 3500 | 0.43 | 0.7 |
| PZT-5H | 7500 | N/A | N/A |

**Table 16.3** PZT-5H coupling matrix in [C/m$^2$]

| 0 | 0 | 0 | 0 | 17.0345 | 0 |
|---|---|---|---|---|---|
| 0 | 0 | 0 | 17.0345 | 0 | 0 |
| −6.62281 | −6.62281 | 23.2403 | 0 | 0 | 0 |

**Table 16.4** PZT-5H relative permittivity matrix [1]

| 1704.4 | 0 | 0 |
|---|---|---|
| 0 | 1704.4 | 0 |
| 0 | 0 | 1433.6 |

**Table 16.5** PZT-5H elasticity matrix in Pascal [Pa]

| 1.27205e+011 | 8.02122e+010 | 8.46702e+010 | 0 | 0 | 0 |
|---|---|---|---|---|---|
| – | 1.27205e+011 | 8.46702e+010 | 0 | 0 | 0 |
| – | – | 1.17436e+011 | 0 | 0 | 0 |
| – | – | – | 2.29885e+010 | 0 | 0 |
| – | – | – | – | 2.29885e+010 | 0 |
| – | – | – | – | – | 2.34742e+010 |

Matrix is symmetric about its diagonal

better approximates how a real motor is fixed. Material properties for the stator (brass), epoxy and piezoelectric ring (PZT-5H) are available in Tables 16.2, 16.3, 16.4, and 16.5.

## Methodology

A PZT-5H 1.190″ (outside diameter) × 0.790″ (inside diameter) × 0.020″ (height) copper plated ring (EBL Products, USA) was chemically etched with the pattern shown in Fig. 16.2. The ring was bonded to a USR30 stator with Loctite 3888 conductive epoxy. Wires were soldered to the exposed side of the crystal breaking out 4 unique phases. It is possible to reduce the number of phases to 2 by poling neighboring sections of the crystal in opposite directions. However, we opted not to pole the segments differently since variations in poling intensity could affect the results. A waveform generator (Rigol DG1022Z, China) was used to create two 90° shifted sine signals. Two additional signals were created by shifting each of the previous sine signals by 180°. The resulting 4 signals were amplified through PA94 (Apex Microtechnology, USA) operational amplifiers set to an inverting configuration powered from a pair of RC125-0.3P and RC125-0.3N (Matsusada, Japan) 300 V power supplies.

The stator with bonded piezo ring was secured onto a fixture designed to be mounted to a standard optical table. Care was taken to ensure a rigid fix and a washer was placed between the stator and mounting screw to ensure uniform stress at the contact boundary. Figure 16.3 shows the stator mounted onto its holder on the optical table.

A digital holography system was setup for stator imaging consisting of a 532 nm 30 mW laser (BWN-532-20E, B&W TEK INC, USA) aimed through an acoustic shutter (AOM-40, IntraAction Corp, USA). The resulting beam is split into beam *A* and beam *B*. Beam *A* is reflected by a mirror fixed to a piezo microactuator powered by a piezo controller (MDT694A,

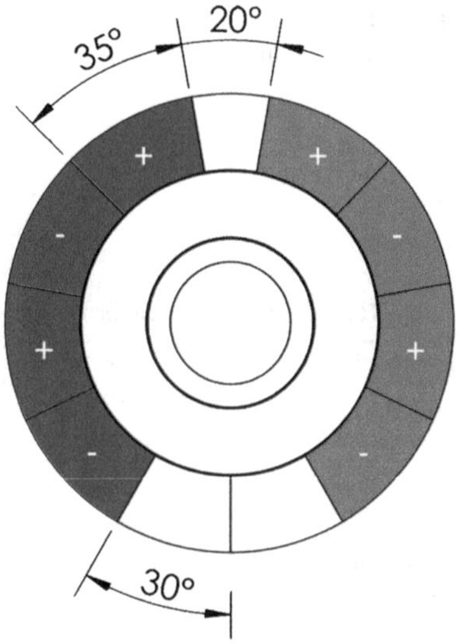

**Fig. 16.2** Electrode patterning of piezoelectric. Blue is sine and red is cosine

**Fig. 16.3** Stator fixed to stand for imaging. Final stage of holography system is visible as is some of the scattered wavefront

Thorlabs, USA). The reflected beam is focused into a fiber through a 20/0.40 lens. Beam B is reflected by a mirror and passes through a 60/0.85 lens to create a wavefront that is shown on the piezo stator afixed to a vertical support on the optical table. The resultant scattered wavefront is picked up by a 55 mm telecentric lens (Computar, USA) and reflected on a wedge. Beam A leaves the optical fiber and is shown on the other side of the same wedge. The combined beam is observed by a camera (Pike F145B, Allied Vision, Germany). The system is a variation of the one used in [18].

## Results

Desired eigenmodes for the constructed stator were determined by manually sweeping the excitation frequency from 100 Hz to 70 KHz at 100 Hz increments while imaging in time-averaged mode. Upon noticing a desired mode, the excitation was fine tuned to where the mode was most discernible. We then captured the mode shape and continued increasing in frequency. During the sweep, excitation voltage was also varied between 10 V and 220 V peak to peak to ensure deformation was within system capabilities. Only modes that would be usable during motor operation are reported. Each captured mode and their corresponding FEM eigenmode is shown in Fig. 16.4a through f. The difference between simulated and measured frequencies

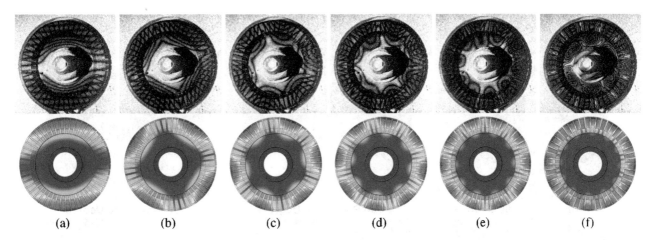

**Fig. 16.4** Time-averaged holography results (top) and eigenfrequency simulation results (bottom) depicting the deformed stator for the first 6 relevant modes. Actual and predicted frequencies [KHz] for each mode are, respectively: (**a**) 1st mode 1.30/3.89 (**b**) 2nd mode 5.21/7.22 (**c**) 3rd mode 13.7/17.14 (**d**) 4th mode 24.56/29.87 (**e**) 5th mode 37.11/44.30 (**f**) 6th mode 50.37/59.74

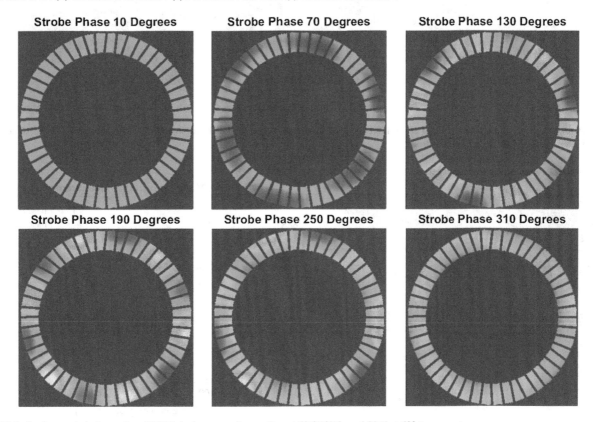

**Fig. 16.5** Stroboscopic holography of PRM during normal operation at 50.37 KHz and 80 V at 60° increments

are partly accounted by manufacturing variations. Variations in frequencies of a few kilohertz were noticed even in between assembled stators (results not reported). All presented work is based on the same stator assembly.

The raw results for stroboscopic holography are shown in Figs. 16.5 and 16.6 for a stator operating at 50.37 KHz. Each image shows the difference in out-of-plane displacement between a reference image, assigned as phase 0, and subsequent strobe phase images. Images are shown for 60° increments in strobe phase despite recording at 10°. Figure 16.6 is an experimental control where motor is being excited by two identical sine signals. Such arrangement does not provide torque to a rotor and should exhibit no rotation.

$$f = A\sin(6\theta + \phi) + \delta \tag{16.1}$$

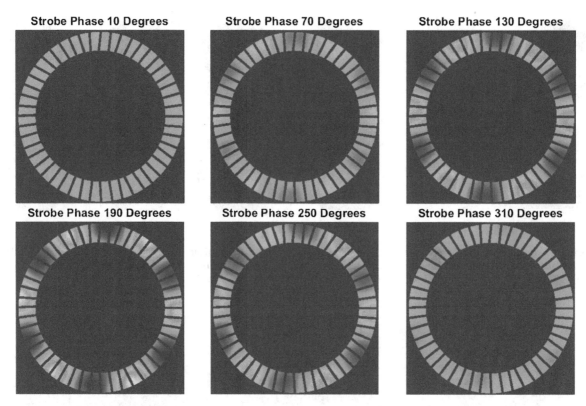

**Fig. 16.6** Stroboscopic holography of PRM at 60° increments being driven by two identical sine waves at 50.37 KHz and 40 V peak to peak. No rotation is visible as expected from the control

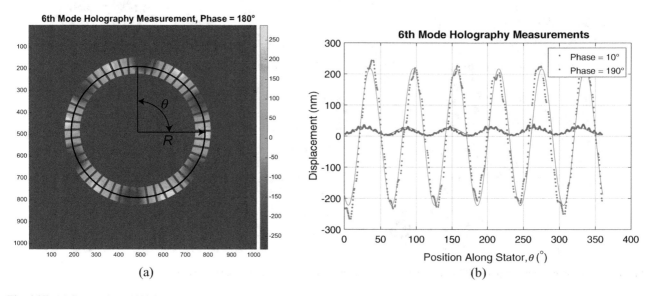

**Fig. 16.7** (**a**) Strobe phase 180° for stator with circle drawn along center of teeth. Colorscale is shown in nm. (**b**) Least square fitting of Eq. (16.1) to raw data for strobe phases 10° and 190°

Quantitative observation of the motion of the peaks in the stator during its deformation is calculated by collecting the displacement values along a circle on the center of stator's teeth as in Fig. 16.7a. The resulting intensity values are shown in Fig. 16.7b. Notice that the number of peaks corresponds to the mode of the stator. We perform a least square fit using Eq. (16.1) for each phase measurement where $A$ is the amplitude, $\theta$ is the angle along the drawn circle, $\phi$ is the relative rotation and $\delta$ is the offset. The resulting $\phi$ for the fitting of each strobe phase is plotted in Fig. 16.8 against its corresponding simulation. The straight lines correspond to the control. Notice that "Measured No Rotation" exhibits some outliers starting at 250°. These are related to the relative amplitude of displacement towards the end of the cycle being too small to ensure

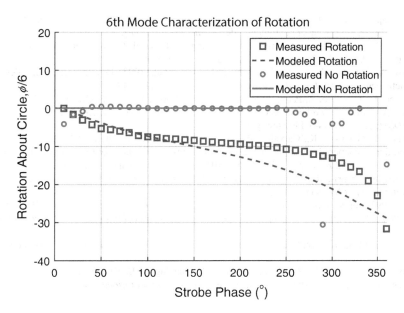

**Fig. 16.8** Comparison of rotary evolution of the peak of the stator's deformation for measured and modeled at excitation frequency of 50.37 KHz. (Red) Control experiment with no rotation. (Blue) Rotating stator

proper fitting. The mismatch between the "Measured Rotation" and "Modeled Rotation" was not predicted and warrants further investigation.

## Conclusion and Future Work

MRI compatible surgical robots are a promising avenue in the treatment of deep brain tumors. Better motors may be needed to improve their compatibility with the scanner. A validated FEM model can be useful in designing more MRI compatible stators. We demonstrated the testing of an FEM model of a PRM stator against a constructed stator through the use of digital holography. The first six desired mode shapes as observed by time-averaged holography were in agreement with simulation. However, a frequency mismatch was observed that should be addressed in future work. Stroboscopic holography was used to capture the out-of-plane motion of the surface waves. Simulation and actual showed a disparity in the rate of peak deformation rotation along the circle that should be further studied.

**Acknowledgements** The authors acknowledge the grants from National Institutes of Health (NIH) (CA166379) and National Science Foundation (NSF) (1428921).

## References

1. Siegel, R. L., Miller, K. D., & Jemal, A. (2017). Cancer statistics, 2017. *CA: A Cancer Journal for Clinicians, 67*, 7–30.
2. Couzin-Frankel, J. (2013). Cancer immunotherapy. *Science, 342*(6165), 1432–1433. https://doi.org/10.1126/science.342.6165.1432
3. Jacobs, L. (2008). Positive margins: The challenge continues for breast surgeons. *Annals of Surgical Oncology, 15*, 1271–1272.
4. Tam, A. L., Lim, H. J., Wistuba, I. I., Tamrazi, A., Kuo, M. D., Ziv, E., et al. (2016). Image-guided biopsy in the era of personalized cancer care: Proceedings from the society of interventional radiology research consensus panel. *Journal of Vascular and Interventional Radiology, 27*, 8.
5. Comeau, R. M., Sadikot, A. F., Fenster, A., & Peters, T. M. (2000). Intraoperative ultrasound for guidance and tissue shift correction in image-guided neurosurgery. *Medical Physics, 27*, 787–800.
6. Stoianovici, D., Kim, C., Srimathveeravalli, G., Sebrecht, P., Petrisor, D., Coleman, J., et al. (2014). MRI-safe robot for endorectal prostate biopsy. *IEEE/ASME Transactions on Mechatronics, 19*, 1289–1299.
7. Yakar, D., Schouten, M. G., Bosboom, D. G., Barentsz, J. O., Scheenen, T. W., & Fütterer, J. J. (2011). Feasibility of a pneumatically actuated MR-compatible robot for transrectal prostate biopsy guidance. *Radiology, 260*, 241–247.
8. Fischer, G. S., Iordachita, I., Csoma, C., Tokuda, J., DiMaio, S. P., Tempany, C. M., et al. (2008). MRI-compatible pneumatic robot for transperineal prostate needle placement. *IEEE/ASME Transactions on Mechatronics, 13*, 295–305.

9. Kokes, R., Lister, K., Gullapalli, R., Zhang, B., MacMillan, A., Richard, H., et al. (2009). Towards a teleoperated needle driver robot with haptic feedback for RFA of breast tumors under continuous MRI. *Medical Image Analysis, 13*, 445–455.
10. Lee, K. H., Fu, K. C. D., Guo, Z., Dong, Z., Leong, M. C., Cheung, C. L., et al. (2018). MR safe robotic manipulator for MRI-guided intra-cardiac catheterization. *IEEE/ASME Transactions on Mechatronics, 23*(2), 586–595.
11. Masamune, K., Kobayashi, E., Masutani, Y., Suzuki, M., Dohi, T., Iseki, H., et al. (1995). Development of an MRI-compatible needle insertion manipulator for stereotactic neurosurgery. *Journal of Image Guided Surgery, 1*, 242–248.
12. Nycz, C. J., Gondokaryono, R., Carvalho, P., Patel, N., Wartenberg, M., Pilitsis, J. G., & Fischer, G. S. (2017). Mechanical validation of an MRI compatible stereotactic neurosurgery robot in preparation for pre-clinical trials. In *2017 IEEE/RSJ International Conference on Intelligent Robots and Systems (IROS)* (pp. 1677–1684). Piscataway: IEEE.
13. Wartenberg, M., Schornak, J., Gandomi, K., Carvalho, P., Nycz, C., Patel, N., et al. (2018). Closed-loop active compensation for needle deflection and target shift during cooperatively controlled robotic needle insertion. *Annals of Biomedical Engineering, 46*, 1582–1594.
14. Carvalho, P. A., Nycz, C. J., Gandomi, K. Y., & Fischer, G. S. (2018). Demonstration and experimental validation of plastic-encased resonant ultrasonic piezoelectric actuator for mri-guided surgical robots. In *ASME 2018 International Mechanical Engineering Congress and Exposition*. New York, NY: American Society of Mechanical Engineers.
15. Sashida, T., & Kenjo, T. (1993). *An introduction to ultrasonic motors*. Clarendon Press/Oxford University Press Inc. ISBN: 0-19-856395-7.
16. Schnars, U., & Jüptner, W. P. (2002). Digital recording and numerical reconstruction of holograms. *Measurement Science and Technology, 13*, R85.
17. Carvalho, P. A., Nycz, C. J., Gandomi, K. Y., & Fischer, G. S. (2018). In *Multiphysics Simulation of an Ultrasonic Piezoelectric Motor COMSOL Conference*.
18. Socorro Hernández-Montes, M., Furlong, C., Rosowski, J. J., Hulli, N., Harrington, E., Cheng, J. T., et al. (2009). Optoelectronic holographic otoscope for measurement of nano-displacements in tympanic membranes *Journal of Biomedical Optics, 14*, 034023.

# Chapter 17
# Digital Volume Correlation: Progress and Challenges

Ante Buljac, Clément Jailin, Arturo Mendoza, Jan Neggers, Thibault Taillandier-Thomas, Amine Bouterf, Benjamin Smaniotto, François Hild, and Stéphane Roux

**Abstract** Digital volume correlation consists in registering series of 3D images of experiments to yield 4D displacement fields. These 4D analyses have been conducted for the last two decades. Some achievements and current challenges are reviewed herein.

**Keywords** DVC · Ex situ tests · In situ tests · Laminography · Tomography

## Introduction

Digital volume correlation (DVC) was introduced 20 years ago [1] and was first applied in biomechanics and then extended to the field of mechanics of materials [2]. Ten years ago, a first review was published by the pioneer of the technique [3]. At that time, results obtained with local registration algorithms were presented. Last year, a second review was published in which the global approaches were also presented [4]. The following discussions summarize some of the achievements over the last two decades and elaborate on some of nowadays challenges.

## Background

DVC is the 3D extension to 2D analyses carried out via digital image correlation (DIC) [5]. It consists in measuring 3D and 4D displacement fields (when image series are analyzed). DVC deals with series of 3D images that can be obtained with various imaging modalities. X-ray tomography is by far the most used modality in the field of biomechanics and mechanics of materials [4]. Stick-like samples are the favored geometry to be visualized. First performed on synchrotron beamlines, computed tomography is more often conducted on lab equipments. X-ray laminography was also used. It allows plate-like specimens to be imaged [6]. Other imaging modalities have also been considered such as optical coherent tomography (OCT) or magnetic resonance imaging (MRI) [4].

Ex situ and in situ tests were then developed to image materials in various states of deformation. Specific testing machines had to be designed to be compatible with the imaging environment [4, 7]. One critical aspect of in situ tests is that the applied loads and corresponding displacements have to be maintained constant during each scan. Furthermore, the environment of the sample can also be controlled in terms of humidity and temperature. These environmental conditions need to be made compatible with the fact that X-rays have to traverse the sample and all the equipment to be acquired by the detector.

## Achievements

For the first decade of DVC developments, only local approaches were implemented and utilized to measure 3D displacement fields [2]. Global approaches, which are mostly based on finite element discretizations of the displacement fields [8], were introduced at the end of the first decade. The second decade saw a significant increase in the number of results reported in the literature with both approaches [4].

---

A. Buljac · C. Jailin · A. Mendoza · J. Neggers · T. Taillandier-Thomas · A. Bouterf · B. Smaniotto · F. Hild (✉) · S. Roux
Laboratoire de Mécanique et Technologie (LMT), ENS Paris-Saclay, CNRS, Université Paris-Saclay, Cachan, France
e-mail: francois.hild@ens-paris-saclay.fr; stephane.roux@ens-paris-saclay.fr

In biomechanics, the failure of various bone structures was analyzed and compared with numerical predictions thanks to DVC analyses [4]. These studies were successful thanks to the multiscale nature of the microstructures that create very high contrast in biological materials imaged via X-ray tomography, MRI and OCT.

Foams, be they ductile or brittle, are a first class of materials that are suited to DVC analyses. Various deformation and degradation mechanisms were revealed and quantified. This is particularly true for indentation tests for which most of the deformation process is hidden under the indentor (i.e., only 3D imaging can be used to analyze the in situ material behavior). The localization mechanisms in granular materials (e.g., sand) were also studied with DVC codes adapted to the grain shapes and their local kinematics.

Localized phenomena such as deformation bands and cracks were studied on various engineering materials. Being a full-field measurement technique, DVC enables such phenomena to be quantified in a very extensive way. Various fields were used to detect and quantify cracks, namely, displacement, strain fields, and gray level residuals. This type of analyses required testing machines to be designed in order to apply, for instance, cyclic loading histories representative of low and high cycle fatigue regimes [7].

Validation and identification of constitutive models have started very recently [4], and they mostly dealt with nonlinear laws written at meso- or microscales. Such approaches are likely to develop more in the decade to come thanks to the achieved reliability and robustness of DVC analyses.

## Challenges

One of the first challenges is related to the suitability of various materials to DVC analyses. In DIC, the speckle patterns are very often created by spraying black and white paints onto sample surfaces. Particles have been added early on to enhance the volume contrast [2]. However, they may alter the behavior of the material of interest. Thanks to various improvements of DVC algorithms, the class of materials that could be analyzed has grown very substantially over the second decade.

Uncertainty quantifications have shown that the displacement uncertainties are generally higher with 3D imaging when compared with DIC procedures, namely, they are of the order of one tenth of a voxel (and one hundredth of a pixel) [4]. For many materials, this means that elastic strains cannot be measured, except when regularized or integrated approaches are implemented [9].

The fact that DVC deals with very large amounts of data becomes very demanding in terms of memory storage, computation time (especially for global analyses) and visualization. This observation calls for new numerical schemes to be implemented. When coupled with numerical simulations at the miscroscale for, say, validation purposes, the latter ones are also challenging. This trend becomes even more severe when the constitutive models are nonlinear with many parameters to calibrate.

The tomography and laminography processes involve acquisitions and 3D reconstructions that can be time consuming. In particular, in lab equipments, the acquisition of high quality images may last one hour or more depending on the selected resolution. Such scan durations do not allow materials with time-dependent behavior to be imaged. In synchrotron facilities, thanks to the brightness of the X-ray beams, these durations can be made significantly smaller but require faster rotation velocities [10]. An alternative route consists in combining DVC and reconstruction steps [11]. With such approaches, the in situ test in no longer interrupted and the radiographs are acquired on the fly [12], thereby allowing mechanical tests to be performed in a few minutes in lab-scale tomographic equipments.

**Acknowledgements** Different parts of the above mentioned examples were funded by Agence Nationale de la Recherche under the grants ANR-10-EQPX-37 (MATMECA), ANR-14-CE07-0034-02 (COMINSIDE), Saint Gobain, SAFRAN Aircraft Engines and SAFRAN Tech. It is a pleasure to acknowledge the support of BPI France within the DICCIT project, and ESRF for MA1006, MI1149, MA1631, MA1932, and ME1366 experiments.

## References

1. Bay, B., Smith, T. S., Fyhrie, D. P., & Saad, M. (1999). Digital volume correlation: Three-dimensional strain mapping using X-ray tomography. *Experimental Mechanics, 39*(3), 217–226.
2. Bornert, M., Chaix, J.-M., Doumalin, P., Dupré, J.-C., Fournel, T., Jeulin, D., Maire, E., Moreaud, M., & Moulinec, H. (2004). Mesure tridimensionnelle de champs cinématiques par imagerie volumique pour l'analyse des matériaux et des matériaux et des structures. *Instrumentation, Mesure, Métrologie, 4*, 43–88.
3. Bay, B. (2008). Methods and applications of digital volume correlation. *Journal of Strain Analysis for Engineering Design, 43*(8), 745–760.
4. Buljac, A., Jailin, C., Mendoza, A., Neggers, J., Taillandier-Thomas, T., Bouterf, A., Smaniotto, B., Hild, F., & Roux, S. (2018). Digital volume correlation: Review of progress and challenges. *Experimental Mechanics, 58*, 661–708.

5. Sutton, M. A., Orteu, J. J., & Schreier, H. (2009). *Image correlation for shape, motion and deformation measurements: Basic concepts, theory and applications*. New York: Springer.
6. Morgeneyer, T. F., Helfen, L., Mubarak, H., & Hild, F. (2013). 3D Digital volume correlation of synchrotron radiation laminography images of ductile crack initiation an initial feasibility study. *Experimental Mechanics, 53*(4), 543–556.
7. Buffière, J.-Y., Maire, E., Adrien, J., Masse, J.-P., & Boller, E. (2010). In situ experiments with X-ray tomography: An attractive tool for experimental mechanics. *Experimental Mechanics, 50*(3), 289–305.
8. Roux, S., Hild, F., Viot, P., & Bernard, D. (2008). Three-dimensional image correlation from X-ray computed tomography of solid foam. *Composites. Part A, Applied Science and Manufacturing, 39*(8), 1253–1265.
9. Bouterf, A., Roux, S., Hild, F., Adrien, J., Maire, E., & Meille, S. (2014). Digital volume correlation applied to X-ray tomography images from spherical indentation tests on lightweight gypsum. *Strain, 50*(5), 444–453.
10. Maire, E., Le Bourlot, C., Adrien, J., Mortensen, A., & Mokso, R. (2016). 20-Hz X-ray tomography during an in situ tensile test. *International Journal of Fracture, 200*(1), 3–12.
11. Leclerc, H., Roux, S., & Hild, F. (2015). Projection savings in CT-based digital volume correlation. *Experimental Mechanics, 55*(1), 275–287.
12. Jailin, C., Bouterf, A., Poncelet, M., & Roux, S. (2017). In situ μCT mechanical tests: Fast 4D mechanical identification. *Experimental Mechanics, 57*(8), 1327–1340.

# Chapter 18
# Development of 3D Shape Measurement Device Using Feature Quantity Type Whole-Space Tabulation Method

Motoharu Fujigaki, Yoshiyuki Kusunoki, and Hideyuki Tanaka

**Abstract** A feature quantity type whole-space tabulation method (F-WSTM) was proposed by authors to make 3D shape measurement devices robust for vibrating. This method makes possible a camera calibration-free 3D shape measurement. Three phase information obtained with three projectors are used to obtain 3D coordinates without any camera parameters. That is, change of lens position does not cause the systematic error. In this method, focusing, zooming, pan and tilt are available anytime. In this paper, a prototype of a 3D shape measurement device using the F-WSTM was developed. The evaluation of the device was performed with an experiment of a shape measurement of a step object.

**Keywords** 3D shape measurement · Feature quantity type whole-space tabulation method (F-WSTM) · Fringe projection method · Camera calibration-free

## Introduction

3D shape measurement using fringe projection method is useful for many fields [1]. In the case of almost all of conventional methods, camera parameters are used to obtain 3D coordinates on the object surface. The method is, however, not robust for vibrating of the measurement device. Especially, the positions of an imaging sensor and lenses are changed easily owing to vibration. It causes some systematic errors.

Authors proposed a feature quantity type whole-space tabulation method (F-WSTM) [2] to overcome this robustness problem. The method extracts the 3D shapes from the phase information of three projectors. The 3D coordinates at a target point on an object are obtained only from the three phase projections at that point, without requiring camera parameters or the point's pixel coordinates in the image taken by the camera. This method makes possible a camera calibration-free 3D shape measurement.

In this paper, a prototype of a 3D shape measurement device using the F-WSTM is developed. An experiment to measure the 3D shape of a step object is performed using the device.

## Principle of F-WSTM

In general, 3D coordinates $(x, y, z)$ can be obtained from an independent set of three values mathematically. Figure 18.1 shows the principle of the F-WSTM. Three projectors $P_A$, $P_B$, and $P_C$ are fixed in a 3D shape measurement device. Each projector is projecting grating pattern onto an object. Projectors $P_A$, $P_B$, and $P_C$ give phases $\phi_A$, $\phi_B$, and $\phi_C$ at point P, respectively. The phases $\phi_A$, $\phi_B$, and $\phi_C$ are obtained by a camera. A set of 3D coordinates $(x, y, z)$ is obtained immediately from a table of feature quantities to 3D coordinates. The table is prepared in advance using reference planes on a calibration process.

Three stable projectors are required for this method. Recently, authors developed a leaner LED device for 3D shape measurement. A compact and stable grating projector can be produced using this device. Figure 18.2 shows an example of

---

M. Fujigaki (✉) · Y. Kusunoki
Graduate School of Engineering, University of Fukui, Fukui, Japan
e-mail: fujigaki@u-fukui.ac.jp; yoshiyuki.kusunoki@asort.co.jp

H. Tanaka
OPTON Co., LTD, Seto, Aichi, Japan
e-mail: tanaka@opton.co.jp

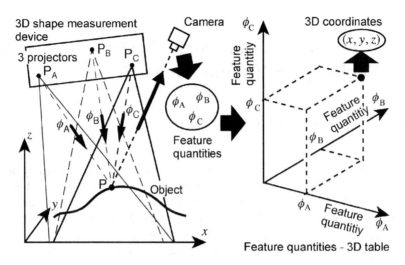

**Fig. 18.1** Principle of F-WSTM

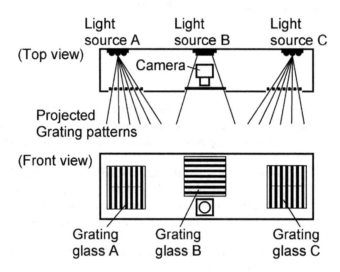

**Fig. 18.2** Example of optical design of 3D shape measurement device using F-WSTM

optical design of 3D shape measurement device using F-WSTM. If any sets of three projected phases are independent in the measurement area, 3D coordinates $(x, y, z)$ can be obtained from the three projected phases.

## Prototype and Experiment

A prototype for confirming the principle of the F-WSTM was constructed as shown in Fig. 18.3. The prototype is installed with three sets of fringe projectors using linear LED devices [3]. Projectors A and C have vertical grating glasses and linear LED devices. They project vertical phase-shifted fringe patterns. Projector B has a horizontal grating grass and linear LED devices. It projects horizontal phase-shifted fringe patterns. A camera is located under the projector B.

Figure 18.4a, b shows a photograph and a drawing of a specimen with 10.0 mm steps, respectively. First, a 3F-3D table (feature quantities and 3D coordinates table) is produced using reference planes. Second, fringe patterns are projected onto the specimen and the images are taken by the camera synchronized with the phase-shifting of the projected fringe pattern.

Figure 18.5 shows the measured result. In this case, the error was around 0.1 mm and the standard deviation is around 0.1 mm.

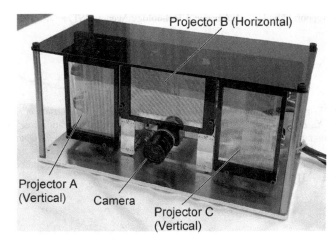

**Fig. 18.3** Prototype. (**a**) Photograph, (**b**) drawing

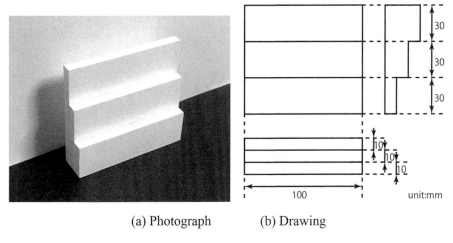

(a) Photograph  (b) Drawing

**Fig. 18.4** Specimen. (**a**) 3D view, (**b**) cross-section

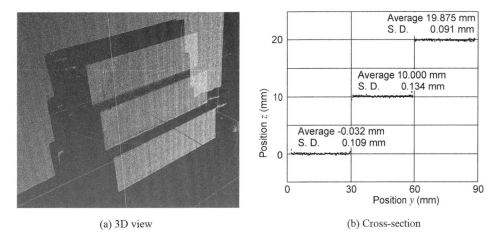

(a) 3D view  (b) Cross-section

**Fig. 18.5** Measured shape. (**a**) 3D view, (**b**) cross-section

## Conclusion

Authors proposed a F-WSTM as a robust 3D shape measurement. A prototype of 3D shape measurement device using the F-WSTM was developed. The experimental result to measure a step object showed the effectiveness of the proposed method.

**Acknowledgements** This study was supported by Japan Science and Technology Agency (JST) as A-STEP Grant AS2915038S.

# References

1. Gorthi, S. S., & Rastogi, P. (2010). Fringe projection techniques: Whither we are? *Optics and Lasers in Engineering, 48*(2), 133–140.
2. Fujigaki, M., Kusunoki, Y., Goto, Y., & Takata, D. (2018). Optical design for camera calibration-free 3D shape measurement using feature quantity type whole-space tabulation method. In *Extended Abstract of International Symposium on Optomechatronic Technology (ISOT2018)*, 2018, pp. 156–157.
3. Fujigaki, M., Oura, Y., Asai, D., & Murata, Y. (2013). High-speed height measurement by a light-source-stepping method using a linear LED array. *Optics Express, 21*(20), 23169–23180.

## Chapter 19
# Temporal Phase Unwrapping for High-Speed Holographic Shape Measurements of Geometrically Discontinuous Objects

Haimi Tang, Payam Razavi, John J. Rosowski, Jeffrey T. Cheng, and Cosme Furlong

**Abstract** We are developing a High-speed Digital Holographic (HDH) system capable of performing near-simultaneous measurements of nanometer-scale displacement and micrometer-scale shape within a fraction of a second during uncontrolled environmental and physiological disturbances, suitable for industrial and life science applications. However, for some applications where time-dependent variations introduce geometrical discontinuities, optical phase measurements become a challenge. In this paper, we present methodologies that overcome these geometrical discontinuities while enabling different levels of measuring resolution. The HDH shape measurements are based on Multiple Wavelength Holographic Interferometry (MWHI). In MWHI the wavelength of the laser is rapidly varied between a series of exposures resulting in micro to millimeter scale synthetic wavelengths. In the holographic recording step, during a single laser tuning ramp, tens of rapid optical phase samplings are acquired, each describing the phase of the object with slightly varied illumination wavelengths (i.e., <0.1 nm). In the holographic reconstruction step, a database consisting of hundreds of phase maps with different synthetic wavelengths is constructed by recovering the shape from any two non-repeated combinations of the phase samplings. A custom temporal phase unwrapping method is developed that resolves and unwraps the depth uncertainties of high-resolution phase maps using the collection of low-fringe density wrapped phase maps from the database that are immune to surface discontinuities. Measurements on National Institute of Standard and Technology (NIST) traceable gauges and discontinuous samples demonstrate the performance of the method. A use of the method in a specific biomedical application is presented.

**Keywords** High-speed holography · Multiple-wavelengths · Shape measurements · Discontinuous phase maps

## Introduction

We developed a High-speed Digital Holographic (HDH) system capable of performing near-simultaneous measurements of nanometer-scale displacement and micrometer-scale shape within hundreds of milliseconds. The shape measurements are based on a multi-wavelength contouring method [1, 2]. By capturing high speed (>10,000 fps) images during a ramp-like variation in the Laser's wavelength, we measure the shape of the sample at different resolutions using different combinations of the varied wavelengths. However, at surface discontinuities with sharp variations in spatial derivatives, unwrapping is challenging and sometimes not possible due to the limited spatial resolution of the high-speed camera. In this paper, we demonstrate the utility of a temporal phase unwrapping strategy utilizing all the wrapped phase samplings with different fringe densities captured during a rapid (i.e., <100 ms) wavelength tuning to resolve high resolution shape phase maps with larger than synthetic wavelength geometric discontinuities. Representative results on a NIST traceable gauge and a human post-mortem eardrum are presented.

---

H. Tang · P. Razavi (✉) · C. Furlong
Center for Holographic Studies and Laser Micro-MechaTronics (CHSLT), Worcester Polytechnic Institute, Worcester, MA, USA

Mechanical Engineering Department, Worcester Polytechnic Institute, Worcester, MA, USA
e-mail: htang3@wpi.edu; prazavi@wpi.edu; cfurlong@wpi.edu

J. J. Rosowski · J. T. Cheng · C. Furlong
Eaton-Peabody Laboratory, Massachusetts Eye and Ear Infirmary, Boston, MA, USA

Department of Otolaryngology, Head and Neck Surgery, Harvard Medical School, Boston, MA, USA
e-mail: john_rosowski@meei.harvard.edu; tao_cheng@meei.harvard.edu; cfurlong@wpi.edu

## Methods

The shape measurement process is described in detail in our previous publications [2]. In this paper, instead of finding the best-resolution measurement by a 2D surface unwrapping algorithm [3], a temporal phase unwrapping [4] method is used. First, wrapped fringe patterns are calculated for all possible phase sampling pairs described in [2], and those fringe pattern maps are sorted based on fringe density. Incremental phase maps ($\Delta\Phi$, defined in [2]) are calculated and maps with phase variation greater than $2\pi$ are rejected. The summation of the remaining incremental phase maps yields an unwrapped phase correlating to shape of the sample. Unlike the method described in [3], the fringe pattern is obtained by multi-wavelength contouring method, and its density is not linearly controlled during the measurements. Therefore, the summation of all the phase difference yields an unwrapped phase map which is proportional to the shape of the sample, but the scaling factor is affected by the exact synthetic wavelength of each wrapped phase. Because it is not feasible to accurately measure the wavelength of the tunable laser during the high-speed wavelength tuning, a calibration process using a spherical National Institute of Standard and Technology (NIST) traceable gauges is applied after each shape measurement to scale the unit of the shape measurement from radians to millimeters.

## Results

Shape measurements of a NIST traceable cylinder gauge were obtained by the HDH method. As shown in Fig. 19.1a, the sample has sharp discontinuities that result in unknown increments of $2\pi$ jumps with respect to maximum synthetic wavelength (Fig. 19.1c), which cannot be easily unwrapped using most conventional 2D spatial phase unwrapping methods (Fig. 19.1c). The use of standard phase unwrapping techniques yields the shape estimates presented in Fig. 19.1d.

These estimates show that the spatial unwrapping approach does not clearly resolve the four discontinuities in the diameter of the test object. However, the temporal phase unwrapping strategy (Fig. 19.1e) successfully resolves the discontinuities of the sample and the unwrapped surface height values are consistent with the NIST nominal dimensions of each of the four observed steps. The same unwrapping strategies are applied to a post-mortem human eardrum as shown in Fig. 19.2. In Fig. 19.2b, 2D spatial phase unwrapping generates unwrapping errors (the sudden jumps in the colormap) at the area where phase quality is low. Figure 19.2c, d show successful reconstructions of the eardrum shape using the temporal unwrapping method.

## Conclusion and Future Work

The proposed temporal phase unwrapping resolves and unwraps the discontinuities of high-resolution phase maps using the information extracted from the analysis of a stack of low fringe density wrapped phase maps. Temporal phase unwrapping is a robust unwrapping strategy with minimal user input during the unwrapping process, which makes temporal phase unwrapping method a good candidate for automatic shape measurement applications. The resolution of the shape measurement can be improved by using a large stack of low fringe density wrapped images. In future work, the application of temporal phase unwrapping will be extended to the displacement measurements obtained by the HDH system to enable an automatic post analysis of both shape and displacement measurements.

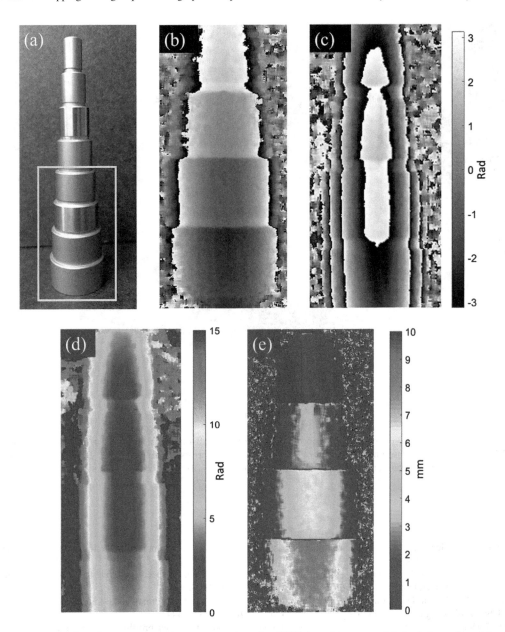

**Fig. 19.1** Shape measurement based on temporal phase unwrapping of a NIST traceable gauge. (**a**) A photography of the NIST traceable cylinder gauge. Yellow box in panel (**a**) shows the measured surface. The diameters of each measured stage from the bottom is 25 ± 0.025 mm, 22.5 ± 0.025 mm, 20 ± 0.025 mm, and 17.75 ± 0.025 mm; (**b**) low-resolution wrapped phase with sufficiently large synthetic wavelength to observe the discontinuities of the sample; (**c**) wrapped phase with highest fringe density (highest resolution) from HDH with clear lack of phase jumps at the location of the discontinuities; (**d**) spatial phase unwrapping [3] of (**c**) illustrating the removal of the sample discontinues; (**e**) scaled shape of the gauge by the proposed temporal phase unwrapping method resolving surface discontinuities (results are scaled into millimeter by calibration and in agreement with the diameter changes of the four stages shown in (**a**)

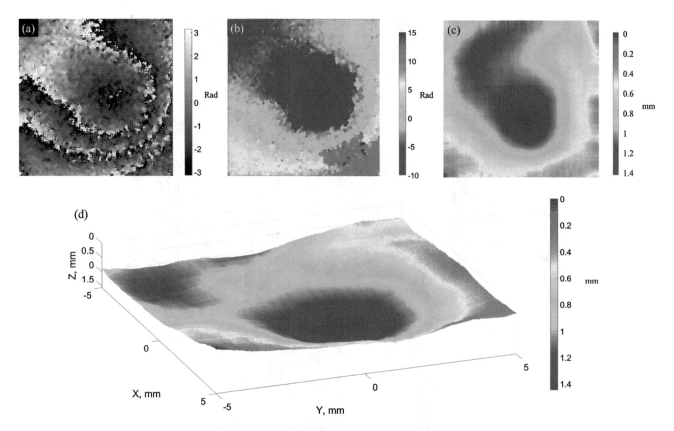

**Fig. 19.2** Shape measurement based on temporal phase unwrapping of a human post-mortem eardrum. (**a**) wrapped phase with highest fringe density from HDH; (**b**) 2D spatial phase unwrapping [2] of (**a**) and unwrapping errors appear due to the limited spatial image resolution. (**c**) temporal phase unwrapping results (results are scaled into millimeter by calibration) (**d**) 3D illustration of temporal phase unwrapping results

# References

1. Wagner, C., Osten, W., & Seebacher, S. (2000). Direct shape measurement by digital wavefront reconstruction and multi-wavelength contouring. *Optical Engineering, 39*(1), 79–85.
2. Razavi, P., Tang, H., Rosowski, J., Furlong, C., & Cheng, J. T. (2018). Combined high-speed holographic shape and full-field displacement measurements of tympanic membrane. *Journal of Biomedical Optics, 24*(3), 031008.
3. Arevalillo Herráez, M., Burton, D. R., Lalor, M. J., & Gdeisat, M. A. (2001). A fast two dimensional phase unwrapping algorithm based on sorting by reliability following a non-continuous path. *Applied Optics, 41*(35), 7437–7444.
4. Saldner, H. O., & Huntley, J. M. (1997). Temporal phase unwrapping: Application to surface profiling of discontinuous objects. *Applied Optics, 36*(13), 2770–2775.

# Chapter 20
# Projection-Based Measurement and Identification

Clément Jailin, Ante Buljac, Amine Bouterf, François Hild, and Stéphane Roux

**Abstract** A recently developed Projection-based Digital Image Correlation (P-DVC) method is here extended to 4D (space and time) displacement field measurement and mechanical identification based on a single radiograph per loading step instead of volumes as in standard DVC methods. Two levels of data reductions are exploited, namely, reduction of the data acquisition (and time) by a factor of 1000 and reduction of the solution space by exploiting model reduction techniques. The analysis of a complete tensile elastoplastic test composed of 127 loading steps performed in 6 min is presented. The 4D displacement field as well as the elastoplastic constitutive law are identified.

**Keywords** Image-based identification · Model reduction · Fast 4D identification · In-situ tomography measurements

## Introduction

Identification and validation of increasingly complex mechanical models is a major concern in experimental solid mechanics. The recent developments of computed tomography coupled with in-situ tests provide extremely rich and non-destructive analyses [1]. In the latter cases, the sample was imaged inside a tomograph, either with interrupted mechanical load or with a continuously evolving loading and on-the-fly acquisitions (as ultra-fast X-ray synchrotron tomography, namely, 20 Hz full scan acquisition for the study of crack propagation [2]). Visualization of fast transformations, crack openings, or unsteady behavior become accessible. Combined with full-field measurements, in-situ tests offer a quantitative basis for identifying a broad range of mechanical behavior.

A now common method to quantitatively measure kinematic data from the reconstructed images is Digital Image Correlation (DIC) in 2D and its 3D extension, Digital Volume Correlation (DVC) [3]. The latter aims at capturing the way a solid deforms between two states from the analysis of the corresponding 3D images. The measured displacement field is then used to calibrate model parameters from inverse procedures (e.g., finite element model updating, virtual fields method). The more numerous the acquisitions (in space and time), the more accurate and sensitive the identification procedure. Those 4D (space-time) analyses (e.g. [4].) always consist of a sequence of three successive inverse problems: (1) volume reconstructions, (2) kinematic measurements from Digital Volume Correlation (DVC), and (3) constitutive law identification (see Fig. 20.1). These approaches, often limited to few scans, are hence extremely dense in space measurements but suffer from sparse time sampling.

A short-cut to the previously described three steps sequence, which is called Projection-based DVC (P-DVC) [5–7], evaluates the full 4D kinematics directly from few selected projections and one single reference volume. The number of radiographs needed for tracking the 3D space plus time changes of the test is thereby reduced from 500 to 1000 down to a single one per time step. With this procedure, the sample is continuously loaded, continuously rotated and regularly imaged by 2D X-ray projections.

---

C. Jailin (✉)
LMT (ENS Paris-Saclay/CNRS/Univ.Paris-Saclay), Cachan, France

Safran Aircraft Engines—SAE, Rond-Point René Ravaud, Réau, France
e-mail: clement.jailin@lmt.ens-cachan.fr

A. Buljac · A. Bouterf · F. Hild · S. Roux
LMT (ENS Paris-Saclay/CNRS/Univ.Paris-Saclay), Cachan, France

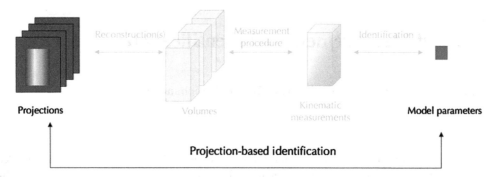

**Fig. 20.1** Dataflow in tomography measurement composed of three inverse problems and its short-cut called Projection-based DVC

## Background

First, tomography consists in reconstructing a 3D volume $f(x)$ for all space locations $x$ from sets of radiographs $p(r, \theta)$ acquired on the detector $r$ at angle $\theta$. Among many tomographic reconstruction algorithms, algebraic schemes are based on the minimization of the following functional with respect to the image voxel intensity

$$\Gamma_{\text{ART}} = \sum_{r,\theta} \|\Pi_\theta [f(x)] - p(r, \theta)\|^2$$

where $\Pi_\theta$ is the projection operator in the $\theta$ direction. Digital Volume Correlation (DVC) [3, 4] is a full field measurement technique for the 4D displacement field that relates a series of 3D images, namely, one volume for the reference state $f(x)$ and few images for the deformed states $g(x, t)$ indexed by time $t$. The DVC procedure (written here with the Eulerian transformation to unify notations, considering the next functional) corresponds to the minimization of the quadratic difference between the reference image corrected by the measured displacement $u(x, t)$ and the images of the deformed state

$$\Gamma_{\text{DVC}} = \sum_{x,t} \|f(x + u(x, t)) - g(x, t)\|^2$$

A kinematic regularization of the displacement field is introduced in global DVC [8] for which the displacement field is expressed on a reduced basis, composed of a set of $N_T$ time functions $\sigma(t)$ and $N_s$ space fields $\boldsymbol{\Phi}(x)$ such that

$$u(x, t) = \sum_{i=1}^{N_T} \sum_{j=1}^{N_s} u_{ij} \sigma_i(t) \boldsymbol{\Phi}_j(x)$$

where $u_{ij}$ are the kinematic unknowns. A general framework for the kinematic bases well suited to mechanical modeling is the framework used in the finite element method. The displacement field is obtained from the minimization of the functional with respect to the space-time degrees of freedom $u_{ij}$ (i.e., the nodal displacements). $N_s N_T$ defines the number of degrees of freedom.

The proposed approach to fast 4D (space and time) measurements is called Projection-based Digital Volume Correlation (P-DVC). The registration consists in minimizing the sum of squared differences between $N_t$ 2D projections $p(r, \theta(t))$ of the deformed configuration, acquired at different times $t$ or angles $\theta(t)$, and loading steps

$$\Gamma_{\text{P-DVC}} = \sum_{r,t} \left\|\Pi_{\theta(t)} [f(x + u(x, t))] - p(r, \theta(t))\right\|^2$$

The procedure makes use of the 3D reference image, $f(x)$, which is reconstructed using classical means (e.g. before the experiment, in a static configuration). As in 3D analyses, a key quantity to validate the procedure is the residual (error) field defined, in P-DVC, as the difference between the projection of the corrected volume and the original projections: $\Pi_{\theta(t)}[f(x + u(x, t))] - p(r, \theta(t))$. Initially, it contains large errors essentially due to the motion. At convergence, these fields indicate what was not captured by the correction model such as incorrect kinematic model assumptions (erroneous choice

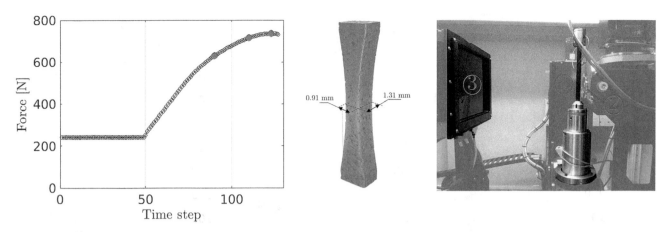

**Fig. 20.2** (Left) Loading for the 127 acquisitions, performed in 6 minutes. The projections shown in Fig. 20.3 are the red dots. (Center) Scanned dog-bone geometry. (Right) In-situ testing device in the tomograph

of $\mathbf{\Phi}(\boldsymbol{x})$ and $\sigma(t)$), noise and reconstruction artifacts. Last, the displacement field is here measured using successive mode identification (PGD measurements) using a greedy algorithm [9, 10].

## Analysis

The present PGD—P-DVC procedure is tested with the analysis of a tensile test on a dog-bone sample made of nodular graphite cast iron until failure. The sample was scanned in the LMT lab-CT (NIS X50+). The test is composed of two main parts (see Fig. 20.2):

- acquisition of a reference static volume (with a small preload) that took 22 min,
- continuous rotation of the sample with 50 projections per full rotation at a rate of one projection every 2 s. One hundred twenty-seven projections were acquired during 2.5 full rotations in 6 min. The first full rotation (i.e., 50 time steps or 100 s) was performed at constant load and was used to quantify the uncertainty. The remaining rotation (starting after 100 s) was carried out with a continuous load change (from 250 to 750 N), as shown in Fig. 20.4, controlled at a constant stroke velocity of 2 μm/s.

After a rigid body motion correction, the full kinematics is identified. The space is regularized by a kinematics (derived from beam theory) driven by 15 rigid sections and trilinear interpolations (i.e., $N_s = 15 \times 6°$ of freedom). The time is regularized by $N_T = 7$ time functions (force measurements, polynomials, sin/cosine). In total, 630 space-time degrees of freedom are finally measured.

The change of the 127 projected residual fields are shown for three selected angles (i.e. acquired at different rotations and loads, see red dots of Fig. 20.2) and are located at the end of the loading sequence, hence correspond to some of the largest strains. The initial and final residuals are shown in Figs. 20.3 and 20.4. A large part of the motion patterns has been erased. Even black parts are appearing in the right image (just before failure) and are due to strain localization. The signal to noise ratio of the 127 residual fields increases from 9.9 to 26.6 dB.

The measured vertical displacement field for each longitudinal section and time is shown in Fig. 20.5(left). From the measured force at each time step, a very simple 1D behavior of the beam is identified. The results of the stress-strain identification for all measurements is shown in Fig. 20.5(right).

## Conclusion

The full space time kinematics of the 6 min experiment was captured with an extension of DVC called P-DVC that uses a model order reduction technique (PGD). Based on highly regularized fields relying on the slender sample geometry as well as a dense sampling in time, this method measures displacement fields from single projections at each time (or load) step of the experiment instead of reconstructed volumes in standard DVC methods. The procedure was tested with an in-situ tensile

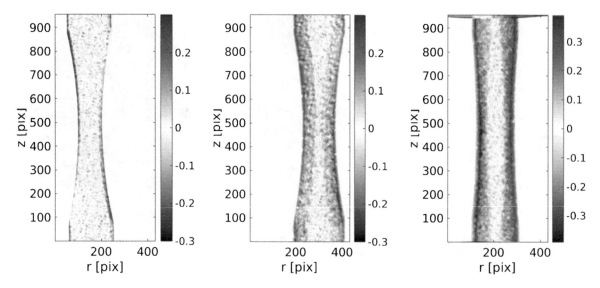

**Fig. 20.3** Three selected initial projected residual fields (see Fig. 20.2). The high positive and negative values are the signature of motions

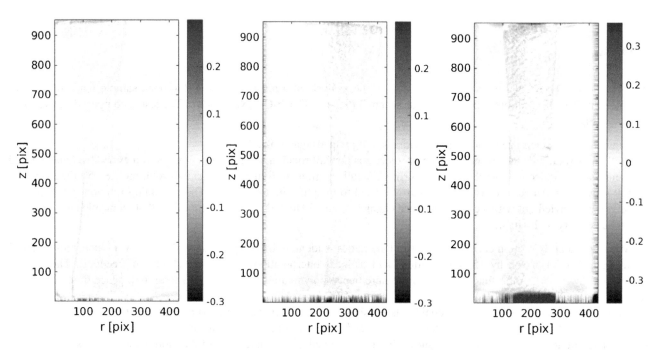

**Fig. 20.4** Three selected projected residual fields, after the kinematic correction. It can be seen that almost all motion patterns have been erased

test on nodular graphite cast iron composed of 127 radiographs with continuous load changes and rotations of the sample until failure. The experiment was carried out in a lab tomograph with an X-ray cone beam source.

The major advantage of PDVC is the important time sampling (and hence temporal resolution). The entire experiment was carried out in 300 s, which is more than two orders of magnitude faster than standard methods. This performance goes together with the benefit of having a continuous (i.e., uninterrupted) loading so that load and rotation can be varied simultaneously. This method could be coupled with 3D DVC analysis and thus benefits both from the space (DVC) and time (P-DVC) huge resolution.

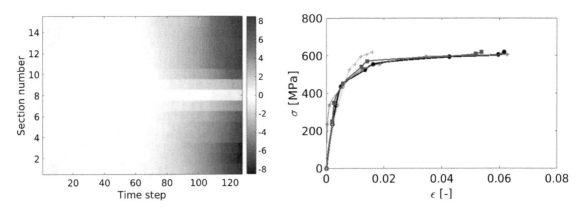

**Fig. 20.5** (Left) Measured vertical displacement field for each 15 sections of the beam as a function of time expressed in voxels (1 voxel ↔ 10.7 μm). (Right) Identified stress-strain behavior

**Acknowledgements** This work has benefited from the support of the French "Agence Nationale de la Recherche" through the "Investissements d'avenir" Program under the reference ANR-10-EQPX-37 MATMECA, and ANR-14-CE07-0034-02 COMINSIDE.

# References

1. Maire, E., Buffière, J. Y., Salvo, L., Blandin, J. J., Ludwig, W., & Létang, J. M. (2001). On the application of X-ray microtomography in the field of materials science. *Advanced Engineering Materials, 3*(8), 539–546.
2. Maire, E., Le Bourlot, C., Adrien, J., Mortensen, A., & Mokso, R. (2016). 20 Hz X-raytomography during an in situ tensile test. *International Journal of Fracture, 200*(1), 3–12.
3. Bay, B. K., Smith, T. S., Fyhrie, D. P., & Saad, M. (1999). Digital volume correlation: Three-dimensional strain mapping using X-ray tomography. *Experimental Mechanics, 39*(3), 217–226.
4. Hild, F., Bouterf, A., Chamoin, L., Leclerc, H., Mathieu, F., Neggers, J., Pled, F., Tomičević, Z., & Roux, S. (2016). Toward 4D mechanical correlation. *Advanced Modeling and Simulation in Engineering Sciences, 3*, 17.
5. Taillandier-Thomas, T., Roux, S., & Hild, F. (2016). A soft route toward 4D tomography. *Physical Review Letters, 117*, 025501.
6. Jailin, C., Bouterf, A., Poncelet, M., & Roux, S. (2017). In situ μ-CT-scan mechanical tests: Fast 4D mechanical identification. *Experimental Mechanics, 57*, 1327–1340.
7. Leclerc, H., Roux, S., & Hild, F. (2015). Projection savings in CT-based digital volume correlation. *Experimental Mechanics, 55*(1), 275–287.
8. Roux, S., Hild, F., Viot, P., & Bernard, D. (2008). Three-dimensional image correlation from X-ray computed tomography of solid foam. *Composites Part A: Applied Science and Manufacturing, 39*(8), 1253–1265.
9. Passieux, J. C., & Périé, J. N. (2012). High resolution digital image correlation using proper generalized decomposition: PGD-DIC. *International Journal for Numerical Methods in Engineering, 92*, 531–550.
10. Gomes Perini, L. A., Passieux, J. C., & Périé, J. N. (2014). A multigrid PGD-based algorithm for volumetric displacement fields measurements. *Strain, 50*, 355–367.